L'Arte Vetraria

The Art of Glass

Volume I of III

L'Arte Vetraria

The Art of Glass

by Antonio Neri

Volume I of III

Translated & Annotated by
Paul Engle

Heiden & Engle

Editorial, Sales, & Customer Service Office:

Heiden & Engle
P.O. Box 451
Hubbardston, MA 01452

Publication Data

Engle, Paul, 1959-
        L'Arte Vetraria : The Art of Glass / Paul Engle. - 1st ed. 2nd Printing
        Includes bibliographical references and index.
        Volume I of a 3 part set:
        ISBN 0-9743529-1-8 (Volume I, pbk)
        ISBN 0-9743529-0-X (3 Volume Set)

Printed in The United States of America

for Lori

CONTENTS

"A fine sieve and dry wood bring honor to the furnace."
-proverb, chap. 8

PREFACE

Undeniably, one of the deepest passions of humanity is the desire to create. Taking raw materials from the earth and working them into the objects of our imagination is at the heart of what makes us tick. The mastery of a single medium is a challenge artists and craftsmen eagerly undertake as a lifelong pursuit. We learn the subtleties and nuances of our chosen materials, reflect upon their properties, imagine what might be possible, and delight in the creation of new beauty. We also take great satisfaction in sharing our knowledge. As novices, we seek to apprentice and learn the secrets of the craft, as peers we make discoveries, challenge each other, and move the discipline forward. Finally as masters, we pass that hard-won precious ability into the hands of a new and eager generation. This passion to create is a key part of our human nature. What we feel about the arts today, our ancestors felt just as strongly four hundred years ago, and four thousand years ago. This is what links us to our counterparts of the distant past, and of the distant future.

L'Arte Vetraria (*The Art of Glass*) by Priest, alchemist, and master 'glass compositioner' Antonio Neri is such an important document for precisely these reasons. It preserves in its pages not only the musings of an inspired glassmaker, but the traditions and technical prowess of an entire community of extraordinary artisans. A community that was at a creative pinnacle four centuries ago when the book was first published. Indeed, a community that had then already flourished for the four centuries previous, and continues to the present day. The Venetian masters at Murano were the Renaissance royalty of glass. Antonio Neri's domain was nearby in Tuscany. Florence was a city very much involved in the craft of glassmaking, as were many of the communities surrounding the preeminent port of Venice. Neri's book opens a window onto this distant past.

Essentially a working notebook, *L'Arte Vetraria* chronicles one man's glassmaking adventures in Florence, Pisa, and Antwerp, where artisans practiced the coveted Venetian techniques. Older accounts of glassmaking have survived from as far back as Babylon, but none like this one. It is for good reason that Neri's 1612 treatise stands out as perhaps the most celebrated book in the history of glass. *L'Arte Vetraria* is a uniquely detailed and methodical treatment of the subject, presented with the intelligence and insight of a master of his craft. It is devoid of the arcane jargon that afflicts the writing of so many of his predecessors and contemporaries. He makes keen observations, and records them clearly and directly. By studying his recipes for glass, we can take a journey back in time, and get a taste of apprenticeship with this man.

In the best of worlds, your translator would be a scholar of seventeenth century Italian, fluent in Florentine vernacular of the late Renaissance, a chemical engineer, an accomplished glass artist, and almost certainly would be of Italian descent. Alas, the reality is that I lack the training, the talent, as well as the genetic material to fulfill these criteria. One qualification I can lay claim to with confidence is a deep and abiding love of glass art. My motivations for undertaking this project spring straight from my heart, and I hope that a large measure of enthusiasm, tempered by patience, respect, and attention to detail is sufficient recompense for my weak credentials. I have worked hard to let Neri's voice come through the translation, and it is my sincere desire to honor this glassmaking priest who put pen to paper in the first years of the 1600s.

For those who choose hot glass as the material with which to explore their creative passions, a very special gift comes with the territory. When we practice the art of glass, we are working side by side in a line of artisans extending literally thousands of years back in time. The tools and techniques we use today have their origins in Mesopotamian hands working fifteen centuries before the birth of Christ. The same tools were being refined a thousand years later by Egyptian royal jewelers during the reign of the Pharaohs. Glassmaking prevailed through the entirety of the Roman Empire, and the rise of Islam saw the work brought to new levels of artistry. Jewish craftsmen drawn from around the Adriatic basin formed the first Venetian glass guilds in the thirteenth century. In Neri's time, a wave of creative and technical achievement radiated out from the glass masters of Murano to Europe and the world. By practicing the art of glass, and learning its history, we get to participate in a fantastic legacy, a legacy that spans many lifetimes and has produced some truly magnificent work.

For those whose passion is glass only after it has cooled, there is an equally proud lineage. Most assuredly as long as there have been glass artisans there have been appreciative collectors. The reason we know the art of glass has been practiced continuously over three and a half millennia is precisely because the custodians of glasswork have protected and preserved their fragile treasures with such devotion and care, sometimes even taking them to the grave. The generosity of benefactors through the ages has both nurtured and advanced the art of glass. In Neri's case, Florentine 'Prince' Don Antonio Medici and the Portuguese noble Sir Emanuel Ximenes of Antwerp apparently sponsored much of the research that made his book possible.

L'Arte Vetraria is the oldest detailed description we have of glass composition technology by a practicing glassmaker. It forms the pedestal upon which rests all subsequent work on the subject. Nearly a half-century passed before the second printing of *L'Arte Vetraria*, yet once it caught on, no less than twenty different versions and translations published before 1800. The subject of our adventures will be the first edition, the original printing of 1612. We will also explore the new preface to the second printing, which appeared in 1661, forty-seven years after Neri's presumed death at the age of thirty-eight, and we will consult the first English translation completed less than a year later in 1662 by British physician and naturalist Christopher Merrett.

Merrett deserves full recognition for producing the world's first translation of *L'Arte Vetraria* and for touching off the landslide of editions that would follow. Had it not been written, Neri's influence on the glass world might well have been much smaller. There can be no doubt that his is a careful and painstaking presentation of Neri's original. Having said that, why embark on another English version? There are several reasons. First, there are notable errors in Merrett that deserve correction. All of them are clearly 'honest' mistakes, (which cannot be said of the editors of some later editions.) Second, it is quite evident that Merrett had a pronounced disdain for Neri, the reasons for which we will probably never know, but it does leave his text unmistakably tainted with his irritation, and leads him to condense and rearrange the work. This too deserves correction. Finally and perhaps most persuasively, is a reason that Merrett had absolutely no control over, the march of time. Baroque English prose is a tough slog for the modern reader, and Neri's work is a little gem that deserves to be read and enjoyed by a wide audience.

After a long period of relative obscurity in the shadow of industry and mass production, the art of hand made glass has entered a new renaissance in popularity, here in the United States as well as around the world. The time seems right for a contemporary English language translation of *L'Arte Vetraria*, not as any great revelation of ancient technique or lost art, but simply as a small step toward connecting us to a truly magnificent heritage.

Paul Engle
Hubbardston
10 March 2003

ACKNOWLEDGEMENTS

First and foremost I would like to recognize the custodians of
the written history of glass; the librarians. It is not an
exaggeration to say that this project would have never gotten
off the ground without the kind attention, patience, and
helpful suggestions of these largely unsung heroes in
institutions around the world. Particularly I would like to
thank Gail Bardhan at the Corning Rakow Museum Library,
and Niki Pollock at the Glasgow University Library. Also
many thanks to the staffs of the rare books departments at
The University of Massachusetts at Amherst, Cornell
University, Yale University, the University of Texas, and the
United States Library of Congress. While important
assistance has been lent from web sites too numerous to
mention, I would like to single out the development and
support teams for the Bibliothèque Nationale de France,
Biblioteca Nazionale di Firenze, Istituto e Museo di Storia
della Scienza, The Medici Archive Project, the Galileo Project,
the Vatican Library, the Tufts University Persius Latin site,
the Alchemy Web Site, Garzanti Linguistica, the Catholic
Encyclopedia site and of course Google. Many thanks also to
Jon Eklund at the Smithsonian Institution, and especially to
Elizabeth and Julia Whitehouse at Whitehouse books, for
unwittingly setting the wheels in motion by selling a rare 1661
second edition of *L'Arte Vetraria* to my wife Lori and me
several years ago. Similarly, many fine authors and editors
have helped to shape my understanding and appreciation of
the late Renaissance, chief among them, the late Dean of
Italian glass historians, Luigi Zecchin. Thanks also to Patrick
McCray, David Freedberg, Philip Ball, and Jacob Burkhardt.
For their invaluable efforts to bring the scarce first edition to
light through contemporary reprints, I thank Rosa Barovier
Mentasti, and also Ferdinando Abbri with Giunti, Neri's
original printer, still going, four hundred years later; a firm
that was already in the book business when Cristopher
Columbus set sail for the new world... simply amazing, or as

the Reverend Priest himself might have quipped: 'Portano la palma'. I would also like to thank John Woodhouse at the University of Oxford, and Kenneth Brown at the University of Calgary for their helpful clarifications, and Emilio Santini, a prince among glass artists, for his discussion and encouragement of the project. Finally, an especially warm thank you goes to my partner Lori Engle, who, by infecting me with her passion for glass and sharing this adventure, has helped make the impossible a reality.

FOREWORD

When I first contemplated an English language translation of
L'Arte Vetraria, I convinced myself that the project could be
completed in six months. I had a wonderful time reading a
wide range of material from existing translations to Florentine
history, to biographies of the Medici family, to the research of
esteemed Venetian glass historian Luigi Zecchin. My six
months flew by, and in the end I had precious little to show
in the way of tangible progress. I decided that I had better get
serious and apply myself. A year and a half later, I still find it
nearly impossible to walk out of a bookstore without
something on the subject of Renaissance history in hand. By
the same token, I have made some real progress, and I am
anxious to share it with you; hence the idea of serializing the
translation. The plan is to release three volumes in paperback,
as they are completed, and then finally an all- inclusive full
color hardcover version. This also gives me the chance to
collect and digest the opinions and critical comments of the
glass and academic communities before publishing a complete
edition.

This is a labor of love, mainly intended for the glass
community. As such, my aim is to bring you Neri as originally
intended, without the numerous changes, embellishments,
omissions and errors extant in previous translations. I hope to
dispel some of the misinformation and mythology that has
plagued the man and his book almost from the start. In
addition to the translation, I have provided the complete
original text verbatim, for those intrepid souls who prefer to
read the Renaissance Florentine vernacular. I have also
included a translation of Luigi Zecchin's research into letters
exchanged between Neri and his friend Emanuel Ximenes,
which helps to put his life in context.

What you think is important to me, and I would very much
like to hear your comments and suggestions. A web site with

current information will be maintained through 2004 (www.heiden-engle.com/neri). I also encourage you to send me a note with any pertinent comments or suggestions. I can be reached electronically at neri@heiden-engle.com and by more traditional means at P.O. Box 451, Hubbardston, Massachusetts, 01452. If I have any regrets about what has been a headlong rush to get this first volume out, it is that there has not been time or resources to secure permissions for pictures. These will just have to wait for the final version.

Finally, I mean no disrespect to accomplished glass artists when I admonish everyone: Please do not endanger yourself, those around you, or the environment by attempting any of these recipes. Hot glass is not to be fooled with unless you know what you are doing. It is a plain fact that the lack of proper equipment and precautions could easily result in poisoning, permanent injury, or death. There is evidence that Antonio Neri himself had a close encounter with chemical toxicity in the spring of 1603, at the Medici glass house in Pisa. Please, please, please, if you are a novice, and you hear the glass-muse calling, contact a reputable organization that can give you appropriate advice on how to get started (GAS - the Glass Art Society, and ISGB - the International Society of Glass Beadmakers are two).

THE ART OF
GLASS
IN SEVEN PARTS

The Reverend Priest
ANTONIO NERI
FLORENTINE.

Reveals marvelous effects, &
teaches his fine secrets of

THE GLASS MELT
& other curiosities.

FOR THE HONORABLE
MR. DON ANTONIO MEDICI.

IN FLORENCE.
Giunti's Print Shop. 1612.
With permission of the Inquisition

L'ARTE
VETRARIA
DISTINTA IN LIBRI SETTE
DEL
R. P. ANTONIO NERI
FIORENTINO.

Ne quali fi fcoprono, effetti marauigliofi , &
Infegnano fegreti bellifsimi ,

DEL VETRO NEL FVOCO
& altre cofe curiofe.
ALL'ILLVST,mo ET ECCELL.mo SIG.
IL SIG. DON ANTONIO MEDICI.

IN FIRENZE.
Nella Stamperia de'Giunti. M. DCXII.
Con licenza de' Superiori.

TO THE HONORABLE
GENTLEMAN
DON ANTONIO MEDICI.

Priest Antonio Neri.

Having spent years of my youth laboring around the glassmaking craft, and having experimented with many fine and marvelous effects, I have now committed these findings to print with the greatest clarity I am capable of, and published them to the world in order to help and delight my fellow artisans. Many of the recipes are my own invention, while others are the work of skilled craftsmen, verified as true.

I shall endeavor to expose and demonstrate the hidden mysteries of this art for the reasons stated above, and if successful, I will be encouraged to publish my work on the subjects of chemistry and medicine as well. I have experimented, on my own and with others, with many impressive effects both reliable and venerable, for no other purpose than to improve my own expertise.

In all consideration, it is my proud duty to dedicate this book to none other than you, most Illustrious Excellency; for you have always been my outstanding patron. You are a gifted leader in this and in all other noble and worthy developments made

ALL'ILLVSTRISSIMO
ET ECCELLENTISS. SIG.
IL SIGNOR
DON ANTONIO MEDICI.

Prete Antonio Neri.

HAVENDO faticato molti anni della giouentù mia, circa l'Arte Vetraria, & hauendo sperimentato in essa molti effetti veri, & marauigliosi, hò di essi compilato vn trattato, con quella maggior chiarezza che per me si è potuto, affine di publicarlo al mondo per giouare, & dilettare per quanto per me si portrà l'intendenti di tal professione, hauendo ritrouato molte cose di mia inuenzione, & alcune altre prouate di valent'huomini, & trouate verissime,

✠ 2 me,

continually in all the arts. This is the essence of a true and
generous Prince. I implore you therefore to accept if not my
work, then my complete devotion to your great merit and
virtue, Excellency. I pray that
God will fill you with happiness.

From Florence
January 6, 1612

me, l'hò volute ftando nafcofte per i fopradetti rifpetti manifeftare, fe haurò con-
feguito cotale intenzione mi fara cagione
per l'innanzi di inanimirmi à publicare il re
fto delle fatiche mie circa alle cofe Chimi-
che , & Mediche , auendo fimilmente , &
nell'vne, & nell'altre fperimentato molti
effetti vtilifsimi, credibili, & mirabili folo à
chi di efsi ne fia per efser' vero conofcitore.
Hò poi giudicato non douere ad altri, che à
V.E. Illuftrifs. dedicare quefto libro sì per effere ella ftata fempre mio fingolar protet-
tore, come etiamdio per effere intendente,
& di quefta, & di ciafcuna altra nobile, &
pregiata cognitione efercitandofi di continuo in tutte quelle difcipline che à vero , &
generofo Principe fi richieggano; fupplico-
la dunque à grandire fe non l'effetto almeno il deuoto animo mio verfo il gran meri-
to,& virtù di V.E. illuftrifs. alla quale prego
da Dio il colmo d'ogni felicità.

DiFirenze il 6. di Gennaio 1611.

TO THE CURIOUS READER

Without a doubt, glass is a true fruit of the art of fire, as it can so closely resemble all kinds of rocks and minerals, yet it is a compound, and made by art. In the fire it fuses together and becomes imperishable. Indeed, like the perfect shining metal gold, the fire refines it, polishes it, and makes it beautiful.

Clearly, its use in drinking vessels and in other utilitarian objects is far more graceful, appealing, and noble than any metal or stone suited for such creations. Beyond the ease and low cost with which it is made, and the fact that it can be made anywhere, glass is more delicate, elegant, and attractive than any material currently known to the world. It is beneficial in the service of herbal distillation, as well as homeopathic healing, not to mention indispensable to the preparation of medicines for man that would be nearly impossible to make without glass. Furthermore, many kinds of vessels and instruments are produced with it; bodies, heads, receivers, distillers, stoppers, retorts, furnace beakers, coils, vials, crucibles, square and round vessels, Florence flasks, and orbs. Countless other types of glass vessels are invented every day to compose and produce elixirs, secret potions, extracts, salts, sulfurs, acids, mercuries, tinctures, elemental separations, all metallic things, and many others that are discovered daily. Also, glass containers are made

AL CVRIOSO LETTORE.

NON è dubbio alcuno, che il vetro è vno de i veri frutti dell'Arte del fuoco, poiche molto ſi aſſomiglia ad ogni ſorte di minerale, & mezzo minerale, quantunque ſia un compoſto, et dall'Arte fatto. Hà fuſione nel fuoco; & permanentia in quello, anzi à guiſa dèl perfetto, e lucido metallo dell'Oro, nel ſuo co ſi affina, puliſce, & faſſi bello. E coſa chiara, che il ſuo vſo in vaſi per bere, & altri commodi per ſeruizio dell'huomo, e molto più gentile, vago, e nobile di qual ſi voglia metallo, ò pietra atta à fare tali lauori, che oltre alla faciatà, e poca ſpeſa con che ſi fà, & in ogni luogo ſi poſſa fare è più dilicato, pulito, & viſtoſo d'ogn'altra materia oggi al mondo nota. Nel ſeruitio poi dell'arte deſtillatoria, & Spagirica è coſi vtile: per non dir neceſſaria ne i medicamenti dell'huomo, che ſaria quaſi impoſſibile ſenza il mezzo del vetro poterſi fare, come di queſto ſi fanno tante ſorti di vaſi, & inſtrumenti, come Boccie, Cappelli, Recipienti, Pellicani, Leuti, Storte, Antenitorij, Serpentine, Fiale, Naſſe, Quadretti, Ampolle, Oui filoſofici, Palle, & infinite altre ſorte di vaſi, che tutto'l giorno s'inuentano per comporre, & fare Eliſiri, Arcani, Quinte eſſentie, Sali, Zolfi, Vitrioli, Mercurij, Tincture, Separationi di Elementi; tutte coſe metalliche, & molte altre, che tutto'l giorno trouano, & fanno oltre alle Acque forte, & Acque Regie; tanto neceſſarie à i partitori, & maeſtri di Zecche dei Principi, per affinare gli Ori, & Argenti, & ridurli alla loro perfezzione. In effetto dal vetro ſi cauano tanti benefizij per ſeruitio dell'huomo, che pare quaſi coſa impoſſibile poterſi fare ſenza il ſuo vſo. è ben ſi conoſce in queſto, ſi come in ogn'altra coſa, la gran prouidentia d'Iddio, che di coſa di tanto biſogno, & vtile all'huomo ha fatto le materie di che ſi compone il Vetro, tanto abbondante in ogni luogo, & Regione, che con molta facilità, ſi può fare per tutto. Il vetro è ancora ornamento grande de i Tempi di Dio, perche di lui oltre à molte altre coſe ſi fanno tante belle vetriate ornate di vaghe pitture, nel le quali i colori metallici ſi aumentano di maniera, & tanto viuamente, che ſembrano tante gioie orientali, & nelle fornacie vetrarie, ſi coloriſcono i vetri di tanti colori, e di tanta bellezza, e perfettione, che non pare ſi poſſa trouare materia a lui ſimile in terra. L'inuentione del vetro, ſi puol credere ſia antichiſſima, perche la Sacra Scrit tura ne i libri di Iob, al Capitolo ventotteſimo, dice Non adequabitur ei aurum, vel vitrum, &c. Il che rende chiara teſtimonianza, che il vetro anticamente foſſe inuentato, poiche San Girolamo (dice Iob)

for nitric and hydrochloric acids, which are so essential for refiners and royal treasurers to purify gold and silver and to bring them to perfection. So many benefits for the service of humanity come from items made of glass, items that seem nearly impossible to make without it.

As is well known in this and all other matters, it is the great providence of God that has put glassmaking materials so abundantly in every place and region, so that something of such great use to humanity can be made with such facility for all. Glass is also a great ornament to God's churches since, among other things, many beautiful windows are made adorned with graceful paintings, in which the metallic colors are so intense and vivid that they seem like so many oriental gems. In the furnaces, glasses of many colors are formed, with so much beauty and perfection that it seems no material like it can be found on earth. The invention of glass, if it can be attributed, is indeed ancient; as the Holy Scripture says in *The Book of Job* 28:17 "gold and glass shall not be equal to it [wisdom]..." which renders clear testimony that glass was invented in antiquity. Saint Jerome held that Job descended from Abraham, and was son of Zanech, who descended from Esau, and therefore was only fifth from Abraham.

There are many who attribute the invention of glass, and perhaps with some reason, to the alchemists; wanting to imitate jewels, they discovered glass. This may not be too far from the truth; as I show clearly, in *Book 5* of the present work, methods to imitate all the jewels, in which I describe the vitrification of stones, that in and of themselves could never be melted or fused. Pliny indicates that glass was discovered by chance, in Syria [now Israel], at the mouth of the Bellus [now Na'aman] river by certain merchants pushed off course at the whim of the sea. They were forced to stop there and set up camp. In order to eat they built fires on the beach, which they fueled with the greatly abundant plant that many call kali [tumbleweed], whose ashes make soda and rocchetta. This burned in the fire, its ashes reacted with the sand, and with the stones prone to vitrify; out trickled glass. An event that illuminated the mind of man on the means and manner to make not only glass, but crystal, crystallino, and many other beautiful things.

Still others hold that in the time of the Roman Emperor Tiberius, a way of making glass malleable was invented; a thing that was subsequently lost, and today is hidden to all. Indeed if such a thing were to be known today, without any doubt it would be more valued than silver or gold for its beauty, and incorruptibility, since glass does not give rise to rust, or taste, or smell, or any other adverse quality. Furthermore, it provides great value to humanity in the use of reading spectacles, and reflecting spheres. Even though the former may also be made of natural rock crystal, also known as mountain crystal, and the later with the alloy known as bronze, a composition made of copper and tin, nevertheless, in the one as well as the other, glass is better suited, not nearly as expensive, more desirable, and more effective. Especially with reflecting spheres. Beyond the difficulty and expense in making them, they do not image as vividly as the glass, and what is worse, in a brief time they tarnish, not reflecting anything at all.

For these and many other reasons, we can easily conclude, glass is one of the most noble things on earth that humanity has today for its use. I have labored in the craft of glassmaking for a long time and seen many things; I am moved to report to the world part of that which I have seen, and done in it. Although the methods to make salts, bollitos [salts for crystal], and frits are well known to many, it still seems to me, that this is a matter requiring clear and distinct treatment, which I do, with observations and diligence. Moreover, if my work is well considered, it will not be judged altogether useless, indeed perhaps even necessary, and enlightening. Besides, with my particular method of extracting salts to make the noblest crystal, the artisan who is diligent in doing things the way I, with clear demonstrations reveal and teach, will make things as worthy, appealing and noble perhaps as anything made today, or that can be made in any other way. In this, and every other matter that I deal with in the present work, the diligent and careful craftsman will find, that I have written and shown the truth, not as I was told, or persuaded by any person whatsoever, but as I actually did, and experienced many times with my own hands.

I have always possessed the resolve to write and to speak the truth. If anyone trying my recipes and methods to make the colors, pastes, and tinctures, does not succeed in making what I have described, do not reject my work or think I have written lies, but reflect upon where you may have erred. Especially those who have never before handled such things, because it is impossible on the first try to have the ability to be a master. That being the case, repeat the work, so that it will continuously improve, and ultimately be perfect, as I describe it. Be warned in particular to give careful consideration to the colors for which exact and determined amounts cannot be given. Indeed, with experience and due practice learn, and with the eye and judgment know, when a glass is colored sufficiently and appropriately for the work at hand. With paste made for the imitation of jewels, in determining the size that you want to make, bear in mind that those to be set in gold with foils, as for rings, and other similar applications should always be more transparent. Those that are also set in gold, but are intended to hang in the air, as earrings and similar items, require a greater charge of color. All of this is

Iob) *effer difcefo da Habraamo, & effer figliuolo di Zanech, defcendente da Efau, & cofi quinto da effo Habraamo. Vogliono molti, & forfe con qualche ragione; che l'inuention del vetro fia ftata trouata dalli Alchimifti: che volendo loro imitare le gioie, trouaffero il vetro, cofa forfe non molto lontana dal vero; poiche come moftro io chiaro nel Quinto Libro della prefente Opera, il mondo di imitare tutte le gioie, nel qual modo fi vede la vetrificatione delle pietre, che per loro ftesse giamai fonderebbono, ne vetrificherebbono. Plinio vuole, che il vetro fuffe trouato à cafo in Soria, alla bocca del fiume Bello, da certi mercanti quiui fpinti da fortuna di mare, & conftretti fermarfi, & per cibarfi nel far fuoco in terra, oue era quantità di quella forte di erba, che molti chiamano Chalì, le cui ceneri fanno la Soda, & Rocchetta, quefta da fuoco abbruciata, & per tal forza vnitofi la fua cenere, & fale con la rena, & pietre atte à vetrificarfi; fi feciono vetro, cofa che illuminò l'intelletto dell'huomo, del modo, & maniera di fare, non folo il vetro, ma il Criftallo, & Criftallino, & tante alter belle cofe, che di effo fi fanno, & di più fi tiene, che al tempo di Tiberio Imperatore, foffe inuentato il modo di fare il Vetro malleabile, cofa poi fmarrita, & oggi occulta del tutto, perche fe tal cofa nota oggi foffe fenza dubbio faria da ftimare per la fua bellezza, & incorruttibilità più dell'Argento, & Oro, fendoche del vetro non nafce ruggine, non fapore, non odore, ne qualità alcuna. Apporta in oltre commodo grande all'huomo nel fuo vfo delli occhiali, & delle fpere, che fe bene l'vno fi puol fare di criftallo naturale detto di Montagna, & l'altro con la meftura detta di acciale, compofitione fatta di rame, & ftagno, nondimeno, & nell'vno, & nell'altro, il vetro è più comodo, di manco fpefa, più vago, & di miglior effetto: maffime nelle fpere, che oltre alla difficultà, & fpefa nel farle, non rapprefentano al viuo, come il vetro, & quell che è peggio: in breue tempo impallidifcono, non rapprefentando cofa alcuna. Onde per quefte, & per molt'altre ragioni: fi puol ben concludere, che il vetro fia vna delle nobil cofe, che habbi oggi l'huomo per fuo vfo in terra. Io hauendo nell'Arte Vetraria lauorato più tempo, et in quella vifto molte cofe, mi fono moffo à dar notitia al mondo parte di quello, che hò vifto, & operato in effa, & fe bene il modo di fare i fali, bolliti, & fritte è noto a molti, tuttauia mi è parfo, che la materia ricerchi trattarfene, come fo io chiaramente, & diftintamente, con alcune offeruationi, & diligentie, che fe faranno ben confiderate, non faranno giudicate del tutto inutili, anzi per auuentura neceffarie, & note a pochi : oltre al modo mio particolare dell'eftrarre i fali per fare vn Cri-*

nearly impossible to teach, but must be left to the judgment of the careful artisan. Keep in mind also, to be meticulous so the materials and colors that you make are well prepared, and well ground. To ensure that materials slated for exquisite work are beyond question, prepare and make the colors personally the same way as I teach, and to that end, you can be confident of good work that will come to fruition happily.

The fire in this art is of notable importance, indeed this is what perfects everything, and without which nothing can be done, therefore give it proportionate consideration. In particular, use hard dry wood, watching from time to time for harmful smoke, which always causes damage, especially in the furnaces, where vessels and crucibles stand open, and the glass becomes imbedded with specks causing imperfections, and notable contamination.

Criſtallo nobiliſsimo, che ſe l'artefice farà nel farlo diligente, ſi come io con chiare demoſtrationi lo paleſo, & inſegno i farà coſa tanto vaga,& nobile, quanto forſe oggi ſi faccia, ò poſſa fare in altra maniera, & in queſto, & in ogn'altra materia, che io tratto nella preſente opera, trouerà il diligente è curioſo operatore, che io h ſcrit to, e moſtrato la verità, non dettami, ò perſuaſami da qual ſi ſia perſona, ma operata, et eſpirimentata aſſai volte con le mie mani, hauendo io ſempe, e hauuto mira di ſcriuere, et di dire la verità. Et ſe alcuno ſperimentando le mie ricette, & modo di fare i colori, paſte, & tincture, non gli riuſciſſe fare quanto io ne ſcriuo: non ſi ſgomenti per queſto, ne creda ch'io habbi ſcritto bugia, ma penſi di hauere erra to in qualche coſa, & maſsime quelli che non hanno mai più manipulato ſimili coſe, perche è impoſsibile, che queſti per la prima volta poſſino eſſere maeſtri: però reiterino l'opera, che ſempre la faranno miglio re, & in vltimo perfetta: ſi come io deſcriuo. Auuertiſco in particolare ad auerſi conſideratione ne i colori de i quali non ſi puol dare certa, & determinata doſi: anzi con l'eſperienza, & pratica ſi deue imparare, & con l'occhio, & giuditio conoſcere, quando vn vetro è colorito à baſtanza, conforme al lauoro perche deue ſeruire, & nelle paſte fatte à imitatione di gioie, conforme alla groſßezza, di che ſi vogliono fare, auuertendo che quelle vanno legate in oro con foglia, come nelli anelli, & altroue, vogliono ſempre eſſere più ſacriche, & quelle che vanno pur legate in Oro, ma deuono ſtare pendente all'aria, come orecchini, & ſimili, vogliono eſßer più cariche, coſe tutte che quaſi è impoſsibile il poterle inſegnare; ma ſi rimette il tutto al giuditio del curioſo operatore. Auuertaſi ancora, & con diligentia, che le materie, e colori ſiano ben preparati, & aſſottigliati di che per eſßer meglio ſicuro, chi vuol fare coſa eſquiſita, prepari, e faccia tutti i colori da ſe medeſimo ſi come io inſegno, perche coſi ſarà certo, l'opera douergli riuſcire felicemente. Il fuoco in queſta Arte è di notabile importanza, anzi quello che perfetiona ogni coſa, & ſenza ilquale niente ſi puol fare: però ſi habbia conſideratione à darlo à proportione, & in particolare con legne forti, & ſecche, guardandoſi dalle loro fumoſità, che ſempre nuoce, & fa danno, maſsimo nelle fornaci: oue i vaſi, & padelle ſtanno aperti, & il vetro poi riceuerebbe imperfettione, & bruttezza notabile. In oltre dico, che ſe l'operatore farà, ò ſi farà diligente, & pratico, & opererà puntualmente, come io deſcriuo, trouerrà verità nella preſente opera, & che io ſolo hò pubblicato, & dato al mondo quanto hò prouato, & eſperimentato. E ſe conoſcerò le mie fatiche eſſer grate al mondo come ſpero: mi inanimirò forſe pubblicare l'altre mie fatiche di tanti anni fatte

in

In closing, I say that the artisan who is diligent, practical, and works step by step, as I describe, will find truth in the present work. He will find that I have published, and given to the world only as much as I have proven and personally verified through experiment.

If the world acknowledges my arduous efforts, as I hope, I will perhaps also be encouraged to publish the experience of my endeavors over many years, working in diverse parts of the world, in the chemical and homeopathic arts. These are matters of nature to which I believe there is no higher calling in the service of humanity.

Known and perfected in ancient

times, men who became experts in these arts were thought to be gods, and were esteemed and revered. I am not boasting to enlarge myself when I say that I have described every last detail clearly and distinctly in this work. Rest assured, that given a bit of experience and practice, as long as you do not purposely foul up, it will be impossible to fail. Therefore take what I have to offer in good course, as I have candidly written this work, first for the glory of God, and then for the experience, benefit and utility of all.

CONTENTS OF THIS WORK

Here will be shown how to extract the salt from Levantine polverino and rocchetta, Spanish soda, ferns, and other plants abundant in Tuscany. Also shown is how to make bollito, which is used in artificial [rock] crystal, and a method to extract these salts chemically for a marvelous crystal. The method to make frit for crystal, crystal-lino, common glass, and [artificial] rock crystal is explained, as are methods to prepare many colors, and in such ways that they will, for the most part, be brighter than ordinary. I show how to make glass with colors of the sky, the blue magpie, golden yellow, garnet, amethyst, sapphire, black velvet, marble, deep red, milk white, peach, Oriental pearl, a light sky blue, and a marvelous aquamarine. I show how to make lead glass in the colors of oriental emerald, topaz, chrysolite [peridot], celestial blue, sapphire, golden yellow, oriental garnet, and other colors. How to make the permanent colors

in diuerse parte del mondo nell'Arte Chimica, & Spagirica, che per seruitio dell'huomo credo non sia maggior cosa nella natura, nota e perfetta nell'antica età, qual fece tener il huomini in quella esperti per Dei, che poi come tali erano tenuti, & reputati. Non mi allargherò dauantaggio perche hauendo io nell'opera descritto ogni particolare, tanto chiaro,& distinto, resto sicuro, che chi non vorrà errare a bella posta sia qualsi impossibile il poterlo fare, hauendone però per prima fatto la esperientia, & practica, Adunque tutto si pigli da me in buona parte, si come io candidamente hò fatto la presente Opera, prima a gloria di Dio, poi a gusto, benefito, et vtile vniuersale.

CONTENVTO DI TVTTA L'OPERA.

SI mostrano i veri modi di cauare il sale del Poluerino, Rocchetta di Leuante, Soda di Spagna, Herba Felcie, & altre herbe abbondanti in Toscana, per far il bollito, che si dice Cristallo artificiale, con vn modo di cauare detti sali chimicamente da far cristallo marauiglioso, & il modo di fare le fritte di Cristallo, Cristallino, Vetro comune, e Cristallo di montagna. Il modo di preparare molti colori, acciò sieno più lampanti,e di farne la maggior parte, & nel vetro il colori celesti, di Gazzera marina, Giallodoro, Granato, Ametisto, Zaffiro, Nero vellutato, Marmorino, Rosso in corpo, Lattimo, Persichino Perla Orientale, di Aierino con vn'acqua marina marauigliosa, & il modo di fare il vetro di piombo in colore di Smeraldo orientale, Topatio, Grisopatio, Celeste, Zaffiro, Giallodoro, Ingranato, & altri colori, & il modo di colorire il cristallo di montagna in color permanente di Rubino, Balascio, Topatio, Opale, & Girasole con il vero modo di fare le paste di tutti i colori, che imiteranno i veri Smeraldi, Topatij, Grisopatij, Zaffiri, Granati, & Acque marine, con vn nuouo modo chimico da farie più dure, e più belle dell'ordinarie, & il modo di fare tutti li smalti da oro di tutti i colori, il Rosicchiero & rosso trasparente, cosa nuoua in Europa, & i modi facilissimi di cauar la Lacca dal Chermisi, Verzino, e Robbia, fiori di Ginestra, Fioralisi, fior cappucci, di borrana, rosolacci, di Mela grana, Rose rosse, incarnate, & d'ogni colore da tutte l'erbe, & fiori, con il modo di fare l'azzurro oltramarino, & altre cose curiose.

for rock crystal in ruby, balas ruby, topaz, opal, and girasol [red opal] is described. I reveal the way to make pastes of many colors that nicely imitate emeralds, topaz, chrysolites, sapphires, garnets, and aquamarines, with a new chemical method to make them more durable, and more attractive than usual. I show the way to make enamels of many colors, and glass in blush-pink rose, and transparent red, which is new in Europe. Also covered is a very easy way to chemically extract the lacquers of brazilwood, goslin weed [madder root], broom flower, blue iris, cabbage flower [golden day lily], borage, wild poppy, pomegranate, red rose, carnation, and every other color found in herbs and flowers. Also, a way to make ultramarine blue, and other curiosities are shown.

~ 1 ~

A New and Secret Method to Extract the Salt from Polverino, Rocchetta and Soda, With Which a Crystal Frit, Called Bollito, Fundamental to the Art of Glassmaking, is Made

Polverino and rocchetta, which come from the Levant [Eastern Mediterranean], and Syria, are the ash of a certain plant, that grows there in abundance. Without a doubt, they make a much whiter salt than Spanish soda. Therefore, when you want to make a crystal of full perfection and beauty, make it with salt extracted from Levantine polverino or rocchetta. Although Spanish soda is richer, and yields more salt, the crystal made with it always inclines to bluishness, it does not have the whiteness and beauty as that made with polverino or rocchetta of the Levant. So here is the way to extract salt perfectly, from both of these, as I have done many times.

Sift the Syrian ash with a fine closely woven screen, so that small pieces are held back and only the rocchetta ash passes through. Crush it up in a mortar made of stone and not of metal, nor with a pestle of iron, which will cause it to take on color and produce results similar to the Spanish soda. Screen it with a fine sieve so that the siftings consist primarily of salt. In buying either of these make sure it is richly salted. This may be determined by touching it with the tongue in order to taste its saltiness; but the surest way of all is to do a test in a crucible and to see if it contains much sand or stones, a thing common in this art and very well known by glass compositioners.

Set up brickwork stoves for copper kettles, like those the

fabric dyers employ. Use more or fewer depending on whether you have occasion to make a greater or smaller quantity of salt. Next, fill the kettles with common clean clear water, and make a fire with dry wood that will not smoke. When the water boils well, throw in the sifted polverino using a fair quantity and proportion to the amount of the water, and continue the fire to keep the water boiling. Continuously stir the bottom with a wooden paddle to ensure that the polverino incorporates with the water, and that all its salt dissolves. Continue to boil the water until it has decreased by a third. Top off the kettles with new water, and boil them until the level decreases by half, at which point you will have a lye solution impregnated with salt.

However, to produce whiter salt and in greater amounts, ahead of the polverino throw into each boiling kettle about 10 [Roman] pounds of the sediment from red wine casks known as tartar. Before hand, roast the tartar just to the point that it turns black. Dissolve it well in the hot water, stirring with the wooden paddle. Then put in the polverino as before. The tartar is the secret way to produce more salt, and crystal which is whiter and of rare beauty. When the water reduces by two thirds, and the salt fully impregnates the lye, relax the fire under the kettles.

Have a number of earthenware pans ready that have first been filled with common water for 6 days; this is to prevent the lye and salt from being absorbed. Next, remove the lye from the kettles with large copper ladles, and

---

A cauare il sale del Poluerino Rocchetta, e Soda con il quale si fa la Fritta del Cristallo detto Bollito, fondamento del arte Vetraria, con vn nuouo, e secreto modo. Cap. I.

IL Poluerino, o Rocchetta, che viene di leuante, & Soria, è cenere di certa herba, che quiui è abbondante, non è dubbio alcuno, che fà il sale più bianco assai, che non fa la soda di Spagna, e però quando si vuol fare vn cristallo di tutta perfezione, e bellezza; si faccia con il sale cauato dal poluerino, o rocchetta di Leuante: perche la soda di Spagna, come più grassa, se bene da più sale, tuttauia il cristallo fatto con il suo sale sempre tira al azzurigno, e non ha quel candore, e bellezza, come quando è fatto con il poluerino, o rocchetta di Leuante. Il modo adunque di cauare il sale perfettamente, e dall'vno, e dall'altro, e l'infrascritto, come più volte ho praticato.

La cenere di Soria si vagli con vaglietto fitto, acciò i pezzeti non passino: ma solo la cenere la rocchetta si pesti in pile di pietra, e non di metallo, perche piglia il suo colore, con pestoni di ferro, & il simile la soda di Spagna, e si vaglino con vaglietto fitto, che in questo consiste il cauarne più, o meno sale. Nel comprare l'vna, e l'altra si auuerta, che sia copiosa di sale: questo si conosce a toccarla con la lingua, per sentire, come sia salata; ma il più sicuro modo di tutti, e farne il saggio in vn coreggiolo & vedere come comporta assai rena, ò tarso, cosa volgare nell'arte, e che i conciatori sanno benissimo.

Si habbino le caldaie di rame murate con suoi fornelli, come

A   quelle

pour it into the pans. Remove the ashes from the kettles, and put all of it into the pans as well. When they are full let them stand undisturbed for 2 days, in which time the ash will all settle to the bottom and the lye will become quite clear. At that point, skim off the lye with a flat copper skimmer so the bottom is not disturbed or stirred up. Put it into other empty earthen pans and leave it undisturbed for 2 days so more sediment will settle to the bottom. The lye will become all the more clear and limpid, and this should be repeated a third time so that you will have the most limpid lye, discharged of all sediment, which will make very fine and perfect salt. Refill the kettles with new water and start them boiling, putting in the 10 pounds of tartar per kettle as before, and polverino as before, and repeat the entire procedure until there is enough material.

To force the lye and extract its salt, first wash the kettles well with clean water and fill them with the refined and clarified lye sited above. As before, bring the kettles to a low boil and keep them topped off with more lye until it thickens and begins to discharge its salt. This should start to happen after a full 24 hours. At the surface of each kettle, you will begin to see white salt that looks like spider webs or white lace. At this point, take a flat ladle perforated with many holes, and keep it in the bottom of the kettle. Collect the salt that falls upon it every now and then, allowing the lye to drain well back into the kettle before putting the salt in basins or earthenware trays, in order for the lye to drain out further. Recover the run-off and return it to the kettle leaving the salt dry. Repeat this process until you have extracted all the salt in the kettles. However, you must closely monitor for when the salt begins to discharge. At that point you should make the fire slow and gentle, because if you made a vigorous fire, the salt would stick to the kettle. If that should happen the salt becomes very potent and always breaks up the kettle, which has happened to me on occasion, so avoid this above all else and exercise great patience and diligence.

Once the salt is in the basin or tray and drained well, collect it, and put it in wooden crates or vats in order to better dry all moisture. The number of days this takes depends on the season in which you make it. Still, as has been demonstrated, the secret of making excellent and plentiful salt resides with the tartar. For every 300 pounds of Levantine ash, I ordinarily extract 80 to 90 pounds of salt. Drain it well, crush it coarsely, and put it in a kiln to dry. Heat it very slowly and with an iron tool known to the furnace artists as a frit rake [reauro], mix the heated salt like frit. Dry it well of all moisture, always paying attention to the kiln that it is not too hot, but moderated. At this point collect the contents of the kiln and crush it very well in a stone mortar. Finally, screen with a fine sieve so that the largest grains allowed to pass are no larger than a grain of wheat. This salt, suitably dried, crushed, and sifted, should be put aside in a place protected from dust. It is now ready for use in making crystal frit whose recipe is presented below as follows.

DELL'ARTE VETRARIA. 3

Per ftrignere dette rannate, & cauarne il fuo fale fi lauino prima bene le caldaie con acqua pulita, & fi empino della fopradetta rannata raffinata, & rifchiarata, come fopra facendo bollire pianamente, e fi attenda a riempiere le caldaie di detta rannata, fino fi veda infpeffare la rannata, che vuol cominciare a buttare il fale, cofa, che fuol feguire in capo di ventiquattro hore in circa, che in fuperficie della caldaia fi comincia a vedere il fale bianco, che pare vna ragna, o tela bianca: all'hora fi habbia vna cazza bucata con più buchi, e fi tenga in fondo la caldaia, & il fale vi cafcherà fopra, e fi caua di quando in quando, lafciando prima bene fcolare la rannata nella caldaia, e fi metta il fale in maftelli, o vero conchette di terra, acciò il ranno fcoli meglio, quale fcolatura, fi recupera, e fi torna nella caldaia, & il fale fi afciuga, e fi continua cofi, fino fi habbi tutto il fale della caldaia, ma bifogna auuertire, quando somincia a buttare il fale di dargli fuoco gentile, & lento, perche fe fi deffi fuoco gagliardo, il fale fi attacheria alla caldaia, & in tal cafo per effer fale potente, rompe fempre la caldaia, cofa a me interuenuta qualche uolta, però fi auuerta quefto fopra ogni cofa, e vi fi vfi gran pazienza, e diligenza: il fale, che è nelle conche, o maftelli, quando è fcolato bene, fi caua, e fi mette in caffe di legno, o tini di legno, per afciugare meglio ogni humidità, che fuccede in più giorni fecondo le ftagioni, in che fi fa, però il fegreto di fare affai fale, & bello confifte nel tartaro, come fopra fi è dimoftrato. Io d'ogni trecento libre di cenere di Leuante per ordinario cauauo da ottanta in nouanta libre di fale; come il fale è bene afciutto, all'hora fi fpezza groffo modo, & fi mette in calcara a fecchare a calore lentifsimo, & con vn'inftrumento di ferro detto riauolo dalli artifti di fornace, fi fpezza, e fi mefcola, come fi la fritta, quando e bene afciutto da ogni humidità, auuertendo fempre, che la calcara non fia troppo calda, ma temperata, all'hora fi caua della calcara, & fi pefta benifsimo in pile di pietra, e fi vaglia con vaglietto piccolo, accioche li maggiori grani, che ne efcano, non pafsino di grandezza il granello del formento.

Quefto fale cofi pefto vagliato, & afciutto fi ferba a parte in

A   2   luogo

~ 2 ~
A Way to Make Crystal Frit, Otherwise Called Bollito

When you want to a make crystal that is beautiful and fully perfect, see that you have the very whitest tarso. At Murano they use pebbles from Tesino [Pavia], a stone abundant in the Ticino River. Tarso, then, is a species of very white hard marble [quartz]. In Tuscany, it is found at the foot of Mount Veruca in Pisa, at Seravezza, at Massa near Carrara, and in the Arno River both above and below Florence. In other places as well, common stone is often recognized, which is seen to have the same qualities as tarso; it is very white and does not have

4   LIBRO PRIMO.
luogo preſeruato dalla poluere per l'vſo di fare la fritta di criſtallo, il cui modo di farla è l'infraſcritto, che ſegue.

Modo di fare la fritta di Criſtallo, altrimenti detto bollito.
Cap.   II.

QVANDO ſi vuole fare vn criſtallo bello, e di tutta perfettione, veggaſi di hauere Tarſo bianchiſsimo. A Murano vſano quocoli del Teſino, pietra abbondante nel fiume Teſino. Il tarſo adunque è vna ſpecie di marmo duro, & bianchiſsimo, che in Toſcana fa a piè della Verucola di Piſa, a Seraueza, & a Maſſa di Carrara, & in nel fiume Arno ſopra, e ſotto di Firenze, & in altri luoghi ancora è pietra aſſai nota, & conoſciuta, ſi auuerta di hauere di quella ſorte di Tarſo, che è bianchiſsimo, che non habbia vene nere, ne giallognola in forma di ruggine, ma, che ſia candido, e puro. Auuertendo, che ogni pietra, che con l'acciaolo, o uero fucile, fa fuoco, e atta a uetrificare, & a fare il vetro, & criſtallo, & tutte quelle pietre, che non fanno fuoco con acciaiolo, o fucile come ſopra non vetrificano mai; il che ſerua per auuiſo per poter conoſcere le pietre, che poſſon traſmutarſi in vetro da quelle, che non ſi poſſono traſmutare.
Queſto Tarſo più bello, & bianco, che ſia poſsibile ſi peſti minutamente in poluere in pile di pietra, & non di bronzo, o altro metallo, acciò non piglino, come piglierebbono il color del metallo, coſa che poi tignerebbe il vetro, e criſtallo, & lo farebbe imperfetto, i peſtoni poi per neceſsità ſieno di ferro, che d'altra materia non potrebbono far l'effetto: poluerizato bene, & fine il tarſo ſi ſtacci con ſtaccio fitto, che tutta l'importanza ſtà, che il tarſo ſia peſtato fine come farina, di maniera, che tutto paſsi per ſtaccio fitto. Pigliſi adunque per eſempio libre dugento di tarſo ſtacciato fine, come ſopra, & di ſale di poluerino peſto, & ſtacciato, come ſopra libre cento trenta, in circa, ſi meſcoli, & vniſca bene ogni coſa inſieme, & coſi vnito, & ben meſcolato ſi metta in la calcara, che per prima ſia ſcaldata bene, perche ſe ſi metteſſe a calcara fredda, la fritta non ſi faria: in principio per vn hora ſi dia fuoco temperato, però ſempre mai con il riauolo ſi meſcoli la fritta, acciò s'incorpori

dark veins, or the yellowish appearance of rust, but is spotless and pure. Take note that any stones that will spark with a piece of steel or strike plate, are apt to vitrify and will make glass and crystal. All those stones that do not make sparks with a piece of steel or striker as above will never vitrify. This serves as advice for being able to distinguish stones that have the ability to transmute their form, from those that cannot be transmuted.

Start with this same tarso, as fair and as white as possible. Grind it finely into powder in stone mortars. Do not use bronze or any other metal for this purpose or the stone will take in the color of the metal, which then would tinge the glass or crystal, and make it imperfect. The pestle must be iron by necessity but at least the other materials will not have the possibility of causing any effect. Pulverize the tarso well and sift with a fine sieve. It is important that the tarso is ground as finely as flour, so that it will all pass through a fine sieve. Now take 200 pounds of the best finely sifted tarso, as before, and about 130 pounds of polverino salt ground and sifted, as before, and stir everything thoroughly and blend it together in its entirety. Once fully mixed, put it in a kiln that has been well heated in advance, because if it were put into a cold kiln the frit would never form. I begin with a moderate fire for an hour, always stirring with the rake and mixing the frit until it is well incorporated and calcined [roasted], then I stoke the fire. Continuously mix the frit well with the rake, which is very important. Keep on doing this without fail for 5 hours, maintaining the strong fire.

DELL'ARTE VETRARIA.

corpori, & ſi calcini bene, poi ſegl'augumenti il fuoco ſempre meſcolando bene la fritta con il riauolo, perche queſto importa aſſai, & queſto modo di fare ſi continui ſempre mai per cinque hore, continuandoil ſempre il fuoco potente.

La calcara e una ſorte di forno calcinatorio, che ſi vſa in tutte le fornacie del vetro, coſa molto nota, e vulgare: il riauolo è ancora lui vno ſtrumento di ferro aſſai lungo, con ilquali ſi agita la fritta continuamente, pure ancora lui ſtrumento aſſai noto nelle fornacie vetrarie; in capo adunque di cinque hore io faceuo cauare la fritta di calcara, la quale quando in detto tempo ha hauuto il fuoco a ragione, & è ſtata agitata bene con il riauolo come ſopra, e ſatta, & ſtagionata, laqual fritta faceuo mettere in luogo aſciutto in palco, & la faceuo coprir bene con vna tela acciò non vi caſcaſſi ſopra poluere, ne immonditie, che in queſto biſogna vſare gran diligenza ſe ſi vuole hauere criſtallo bello, la fritta quando è fatta con le diligenze ſopradette, viene bianca, e candida, come vna neue del cielo. Quando il tarſo è magro, ſe li dia dieci libre di ſale di più alla doſi detta in circa: imperò alla prima fritta ſi fa ſempre ſperienza da i pratichi conciatori di metterne in vno correggiuolo, e queſto meſſo in vna padella di vetro pulito vedere, ſe puliſce bene, & preſto, & di coſi ſi vede, ſe la fritta è tenera, o dura, & all'hora, o ſi accreſce, o ſi minuiſce la doſi del ſale. Queſta fritta di criſtallo, come ſopra ſi è detto ſi tiene in luogo aſciutto, oue non ſia punto di humidità, perche ne i terreni, & luoghi humidi la fritta di criſtallo patirebbe aſſai: poiche il ſuo ſale ſi ſciorebbe, & andrebbe in acqua, & rimarrebbe il tarſo ſolo, quale per ſe non vetrificherebbe, ne anco ſi bagna queſta fritta, come ſi fanno l'altre, e quando ſi laſſa ſtare fatta tre, o quattro meſi è molto meglio per mettere in padella, & più preſto puliſce. Queſto è il modo di far la fritta di criſtallo con ſue doſi, & circoſtanze, come io hò fatto più volte.

Altro

The frit kiln is a kind of calcining furnace that is used in all of the glass houses; it is very well known and commonplace. The rake is a very long iron instrument with which the frit is agitated continuously, and also is a well known tool in the glass houses. At the end of 5 hours, I then take the frit out of the kiln, which in that time has had a radiant fire, and has been mixed well with the rake, and made and seasoned as described. I store the frit in a dry place and I cover it well with a cloth so that no dust or dirt will fall upon it. If you want to have fine crystal, then in this you must exercise great diligence; when the frit is made with this careful attention, it will be white and pure like snow from heaven. When the tarso is lean, you must add about 10 pounds more salt then the prescribed dosage. For this reason, before the frit is finished, a practice often used by experienced glass compositioner is to put some in a crucible and add it to a [hot] furnace pot of clean glass to see if it clarifies nicely, and quickly. This way he sees if the frit is soft or hard, and whether to increase or decrease the dosage of salt.

Keep this crystal frit, as said before, in a dry place where it is not exposed to moisture, because in regions and places that are humid it will suffer greatly. Soon the salt would soak up water and would dissolve away; all that would remain is the tarso alone which would not vitrify. Furthermore, this frit should not be washed the way others are. When it is allowed to rest for 3 or 4 months it is in much better shape to put in crucibles [in the furnace], and will clarify sooner. This is the way to make crystal frit with its dosages and circumstances as I have made many times.

~ 3 ~
Another Method to Extract the Salt from Polverino, Which Makes a Crystal as Beautiful and Clear as Rock Crystal. This New Method is My Own Invention

Take well-sifted Levantine polverino and put it in large glass chamber pots that are well luted [coated with refractory clay] on the bottom. Set them in ash or sand on the furnace, and give them a slow fire. At first, fill them with common water, and keep them on a moderated fire for several hours. Leave them until half the water has evaporated, then cool the furnace and decant the water into glazed earthenware trays. Replace it with new water poured over the remainder of the polverino and boil as before. Repeat this process until all the salt has been extracted by the water. This is determined by checking when the water is no longer salty to the taste, and no longer tinted with color to the eye.

Take as much of this lye as is desired, filter it, and put it aside in glazed trays to rest for 4 to 6 days so that it will leave most of its sediment. Then return it to the filter again, as such it will be purified and separated from most of its sediment. At this point, evaporate the lye in clay-coated chamber pots, on a furnace, set in ash or sand on a slow fire. Finally, as all the material dries up, make sure than the fire is at its slowest so that the salt will not be burned and wasted. When it is well dried, remove the chamber pots in order to see if they became cracked on the bottom. This is apt to occur often, in which case move the salt into other good chamber pots, also coated with clay on the bottom, and cover with pure clear common water.

Again, set them on the furnace in ash or sand on a slow fire. Always evaporate one-eighth of the water, then cool the furnace and empty this water, now fully impregnated with salt, into glazed earthenware trays. Allow it to settle for 24 hours. Filter it carefully so the salt is free of the rest of the sediment and dregs. On a slow fire evaporate this filtered lye in chamber pots and finish with the fire slower still, so that the salt is not burned. The salt is again poured into chamber pots and dissolved with common water as before. Repeat this work, all in all, until the salt does not leave

6   LIBRO PRIMO.

Altro modo di cauare il sale del poluerino, che fa il cristallo tanto bello, & chiaro quanto il cristallo di montagna, modo nuouo da me inuentato.   Cap. III.

PIGLISI il poluerino di Leuante bene stacciato, & si metta in orinali grandi di vetro lutati in fondo a cenere, o rena in fornelli, se li dia fuoco lento, hauendoli prima pieni di acqua comune, dandoli fuoco temperato per più hore nel fornello, & si lascino tanto, che suapori la metà dell'acqua, freddato il fornello, si decanti pianamente l'acqua in catinelle di terra inuetriate, rimettendo nuoua acqua sopra le residenze del poluerino, e si bolla come sopra. Questo si reiteri fino l'acqua habbi cauato tutto il sale, che si conosce quando al gusto l'acqua non è più salata, & a l'occhio non è più carica di colore; si habbi di questa liscia quella quantità si vuole, e si feltrino queste liscie, & feltrate si lascino stare in catinelle vetriate a posare per quattro, o sei giorni, che così lasceranno vna gran parte del lor terrestreità, poi si tornino a feltrare di nuo uo, & cosi saranno purificate, & separate da vna gran parte della lor terrestreità, all'hora si mettino queste liscie a suaporare in orinale di vetro lutati in fondo in fornelli, in cenere, o rena a fuoco lento, & in vltimo quando si asciuga la materia, si auuertisca, che il fuoco sia lentissimo acciò non si abbruci, & guasti il sale, quale asciuto bene, si caui delli orinali per vedere se fussero rotti nel fondo, che suole succedere spesso, nel qual caso si metta detto sale in altri orinali buoni, pur lutati in fondo, & per sopra si empino d'acqua comune pura, & chiara, & in fornello in cenere, o rena a fuoco lento: sempre mai si faccia suaporare vn'ottauo di detta acqua, poi freddo il fornello si voti quest'acqua piena, e pregna di sale in catinelle di terra inuetriate, & come l'acqua e posata ventiquattro hore, si feltri con diligenza, che il sale lascierà dell'altra terrestreità, e feccie, si suapori questa liscia feltrata in orinali a fuoco lento, & in vltimo più lento, acciò il sale non s'abbruciassi, quale sale di nuouo si torni in orinali, & con acqua comune a soluere, come sopra in tutto, e per tutto, reiterando quest'opera, fino, che il sale non lascia dopo di se più terreistreità, ne
                                        fecce

behind any more sediment or dregs. You will have a most pure and perfect salt, which when aged with white tarso, makes frit, as said before, and in the end will make a crystal that is of great beauty, whiteness, and splendor that will surpass rock crystal and even oriental crystal. This work must be done in glass vessels and not copper, as used before, because the salt always draws in the color of the metal and therefore always becomes greenish. While this is somewhat laborious work, and only makes a small amount of salt, it will however make a crystal worthy of any great Prince and will be suitable to make any sort of vessel or fine piece. This was my invention, which I have tested many times with blessed success, to my great satisfaction.

~ 4 ~

A Caution About Golden Yellow in Crystal

You should be warned about mixing polverino salt with tartar salt in order to make beautiful white crystal as described in the beginning. Because frit made that way with that salt is not good for making golden yellow, and in such case you cannot make this color, but can make all the other colors. In order to make a golden yellow you must make frit with salt extracted solely from polverino, and purified as described above, otherwise your yellow will not emerge.

A Method to Make Salt from a Plant Known as the Fern, Which Makes a Very Nice Crystal

In Pisa, I gained experience with the ash of a plant called the fern that grows in great abundance in Tuscany. Cut this herb from the ground when it is green, between the end of the month of May and mid June. The moon should be waxing, and close to its opposition with the sun, because at this point the plant is in its perfection, and gives a lot of salt, more than it would at other times, and of better nature, strength, and whiteness. When you leave it to dry upon the ground uncut, it gives little salt, and is of little good. So cut the plants from the land, as stated, pile them up and leave them to wither.

At this point, burn them thoroughly to leave their ash. From this ash, with the rules, observations, and diligence prescribed for Levantine polverino salt, extract a good pure salt. With it and well-sifted prime tarso I have made frit; frit that melted very well in the crucible, which composed a beautiful crystal for me. It was much more pliable than ordinary crystal; it was very strong and very flexible when stretched into thin threads, as I pulled it, which ordinary crystal is not. This frit can be given a wonderful golden

8 LIBRO PRIMO.

fua eenere. Da quefta cenere con le regole, offeruationi, & diligenze dette di fopra nel fale del poluerino di Leuante fe ne caua vn fale purificato, & buono: del quale io feci fritta con tarfo bello, e ben ftacciato, la qual fritta in padella colo beniffimo, & mi dette vn criftallo bello, & molto più dolce del criftallo ordinario, poiche haueua affai neruo, & fi piegaua molto più, che non fa il criftallo ordinario, tirifi in fili fottili, come lo feci tirare, & a quefta fritta fe il può dare il colore del giallo d'oro ftupendo; auuertendo non vi fia dentro fale di tartaro, come fopra fi è aúuertito, perche ne anco in quefta verria il giallo d'oro, & il giallo d'oro, che fi da a quefto criftallo viene affai più bello, & vago, che non fa nel criftallo fatto con il fale di poluerino di Leuante, & di quefto criftallo fe ne può fare ogni forte di lauoro, come dell'altro.

Modo di fare vn'altro fale, che farà vn criftallo marauigliofo
& ftupendo. Cap. VI.

FACCISI cenere con il modo fopradetto de i gufci, & gambe di faue fecche la ftate, quando i contadini hanno battuto, & cauatione le faue, da quefte cenere con le regole, & diligenze dette nel fal del poluerino di Leuante, fe ne caui il fuo fale, quale farà marauigliofo, del quale fattone fritta con tarfo bianco, & ben ftacciato, come fopra s'è detto diffufamente, fi farà vna fritta nobiliffima, laquale in padella farà vn criftallo di tutta bellezza; il medefimo fi farà dalle ceneri dei cauoli, del Rouo, cioé fpino, che fa le more, & da fagginali ancora, & da i giunchi, & cannuccie de'paduli, & da molte altre herbe, che daranno il lor fale, con il quale facendo fritte al folito, fi faranno criftalli belliffimi, come ogni fpirito gentile, & curiofo potrà con l'efperienza prouare, perche con l'efperienza fi troua, & impara più affai, che non fi fà con lungo ftudiare.

Modo

yellow color provided there is no tartar salt within, as described in the caution, because then golden yellow will not emerge. This crystal is given to a golden yellow that is far more beautiful and pleasant than can be achieved in crystal made with Levantine polverino salt, and with this crystal unlike the other, every kind of job can be done.

~ 6 ~
A Method to Make Another Salt that will Make a Marvelous and Wonderful Crystal

Set about making ash in the way previously described, however use the husks and stalks of fava beans after the farmhands have thrashed and shelled them. With the rules and diligence prescribed for the Levantine polverino salt, extract the salt from this ash, which will be marvelous, and from which a frit can be made using well-sifted white tarso, as is described throughout this work. A very noble frit will result, which in the crucible will make a crystal of all beauty. The same may be made from the ashes of cabbages, or a thorn bush that bears small fruit, called the blackberry, even from millet, rush, marsh reeds, and from many other plants that will relinquish their salt. Use these to make frit as per usual, and you will make the most beautiful crystals. Every noble and curious spirit should try them for the experience,

Sale, che farà vno Criſtallo aſſai bello.   Cap. VII.

CAVISI il ſale della calcina, che ſerue per murare, & queſto ſale purificato ſi meſcoli con il ſale del poluerino di Leuante ordinario a ragione di libre dua per cento, cioè libre dua di ſale di calcina, & libre cento di ſale di poluerino purificato, & ben fatto, come ſopra ſi dice di queſto ſale coſi meſcolato ſi faccia fritta all'ordinario, & ſi metta in padella a pulire: come ſi dirà auanti nel modo di fare il criſtallo criſtallino, & vetro comune, che coſi s'ha uerà vn criſtallo aſſai vago, e bello.

Modo di fare la fritta ordinaria, cioè di Poluerino, di Rocchetta, & di ſoda di Spagna.   Cap. VIII.

LA fritta nõ è altro, che vna calcinatione de i materiali, che fanno il vetro, e ſe bene ſenza calcinatione fonderebbono, e farebbono vetro tuttauia, queſto ſuccederebbe cõ vna lunghezza di tempo, & faſtidio grande, e però è ſtata troua ta tal calcinatione di fare, & calcinare la fritta nel fornello detto calcara, che quando è ben calcinata, & la doſi de i materiali ſia giuſta, conforme alla bontà delle ſode, fonde preſtiſſimo in padella, & puliſce a marauiglia. La fritta, che è fatta di poluerino, farà il vetro bianco, però farà vetro ordinario, la fritta fatta di rocchetta pur di Leuante fa il vetro belliſſimo, che ſi dice criſtallino, la Soda di Spagna, ſe bene è più graſſa delle ſopradette per ordinario, tuttauia non fa il vetro coſi bianco, & bello, come la di Leuante: perche ſempre tira vn poco all'azzuringno. Adunque per farla, il poluerino ſi ſtacci per ſtaccio fitto, i minuzzoli, che non paſſano ſi peſtino in pile di pietra, & non di metallo, acciò non piglino del ſuo colore, il ſimile ſi faccia alla Rocchetta, & alla ſoda; però ciaſcheduna da per ſe: & in effetto operare, che ſieno ben peſte, & ſtacciate per ſtaccio fitto, che come dice il volgar prouerbio nell'arte vetraria, Staccio fitto , & legna ſecca fann' honore alla fornacie. Adunque qual ſi ſia di queſte ſode, per eſempio libre cento di ſoda vuole ordinariamente libre ottantacinque in nouanta di tarſo, quale, come

B   ſopra

because through experience one can discover and learn much more than could ever be accomplished through long study.

~ 7 ~
A Salt that Will Make a Very Beautiful Crystal

Take lime salt [calcium oxide], which is used for building. Purify this salt and mix it with ordinary Levantine polverino salt in the proportion of 2 pounds per 100, which is 2 pounds of lime salt, to every 100 pounds of polverino salt, purified and well made, as previously described. With this salt mixture, you can make yourself ordinary frit, and put it in the crucible to clarify. I will refer to this frit from now on in the recipes for crystallino, crystal, and common glass. This way you will have crystal quite delightful and beautiful.

~ 8 ~
A Way to Make Ordinary Frit with Polverino, Rocchetta, and Spanish Barilla

Frit is nothing other than a calcination of the materials that make up glass, and although one could melt them without calcination, and would make glass, this would only succeed after a protracted period of time and a great amount of trouble. Consequently, the calcination process was developed to make and to calcine frit in an oven called a frit kiln. When it is well calcined and the doses of the materials are right, and conform to the quality of the barilla [soda], it fuses readily in crucibles and it clarifies marvelously. Frit that is made of polverino will make white glass, but it will be ordinary glass, the frit made of rocchetta, also from the Levant, makes a more attractive glass called crystallino. Barilla of Spain is richer than the polverino for ordinary glass; nevertheless, it does not make the glass quite as white and beautiful as that of the Levant because it always tends to have a bluish cast.

So in order to make it [polverino frit], sift the polverino through a fine sieve. Crush the small pieces that do not pass through in a stone mortar, not metal, so that it will not take on any color. Do the same thing for the rocchetta and barilla, but do each one by itself, and continue to work them until they are well ground and sifted in a fine sieve. As the common proverb of the art of glassmaking says: a fine sieve and dry wood bring honor to the furnace. Then with any of these sodas, 100 pounds of soda ordinarily requires 85 to 90 pounds of tarso, for example. As previously described for making crystal, this should be finely ground in a stone mortar only, and not in any other kind. Now pass it through a fine sieve using more or less depending on the quality of the soda and its richness. You must always perform an assay of how much to put in, as is well known in the art.

Now sand, especially that of Tuscany, which is found in the Arno River valley is much richer and has much more salt already in it than tarso. So never add more than 6 to 8 pounds [salt] per 100 [sand]. Before it can be used, this sand must be well washed of all its extraneous sediment and sifted. It will then make a nice white glass, but tarso always makes the most beautiful of any sand found in Tuscany. Procure the required dose of the sand or tarso as described above, first mix it together and incorporate it very well with the soda or well-sifted polverino salt. Then put it in a hot frit kiln and spread it well. Begin mixing it and blending it with the rake in the kiln, so it calcines well. Continue for 5 hours, always maintaining the fire, until it begins to form lumps and grow into little pieces, like nuts. When the fire has been maintained as described, and the frit has been well worked with the rake for the indicated space of 5 hours, the frit will be prepared in full perfection. If you want to see if it is well made then take a little out. As it cools if it lightens to a yellowish white, then it is finished. However, calcining it more than 5 hours is not bad, because the longer it is worked the better it is calcined, and the quicker it fuses in the crucible. Being in the frit kiln a little longer consumes and purges the yellowishness and scum that it has within it and the glass becomes cleaner and purer. Remove the frit from the kiln when it is hot and glowing, pour 3 or 4 buckets of fresh water over it then put it in the cellar in a damp, cool place.

Take the leftover sludge from making salt, as previously described [chapters 1, 3], and put it in the same pans used for the polverino lye. Pour common water over it, and position basins under the pans so that they receive the water, which little by little drips through the residual sludge, and becomes a much stronger lye. Store this pure and clear lye separately and every now and then, sprinkle it over the frit. When the frit is piled up in a moist place and sprinkled with the lye for the space of 2 or 3 months or more, the longer the better, it will solidify together like rock such that you must break it up with a pickaxe. Now when it is in the crucible it will melt wonderfully in a short time and make the whitest glass; almost like crystal. This effect is caused by the lye, which leaves behind its salt. When lye is not available, sprinkle it with common water. Even if it does not have as great an effect as the lye, it nevertheless does work well, and it helps a lot to facilitate fusion. Therefore frit should always sit for some months, which when made this way always requires less firewood to be consumed, and makes the glass smooth and pliable to work.

~ 9 ~

To Make a Fully Perfect Crystal

You should start with crystal made step by step as was shown at the beginning of this book. This you should put into crucibles in the furnace, where there are no crucibles of colors, as the fumes of the metals with which most colors are made turn the crystal ugly and pale. But in order to make it become beautiful and dazzling white you must parcel a quantity of manganese into the frit, while it is in the crucibles in the furnace. The amount that is distributed to the crucibles depends on whether they are larger or smaller, and this is a matter that rests in the practice of the glass compositioners as is their responsibility. This is manganese of Piedmont only, and of no other place, well prepared as will be described. The furnace should have dry wood, hard wood of oak because soft wood tinges the furnaces and does no good. Stoke it steadily and continuously so that the flame is always clear, and there is never any smoke, which is very important in order to make a beautiful crystal.

After a while, when the glass is well fused, take it out of the crucibles and throw it into large earthenware pans or clean sturdy wooden tubs filled with fresh water. This step of throwing the glass into water has the effect of causing the water to remove a kind of salt called Alkali salt [glass gall], which ruins the crystal and makes it dark and cloudy. So while it is still being worked let the glass spit out this salt, a substance quite foul, then return it to clean crucibles. Carry out this flinging into water repeatedly as necessary. In order to separate the crystal from all its [alkali] salt, this should be repeated to the satisfaction of the furnace compositioner.

Then leave it to boil for 4 to 6 days and, to the extent possible avoid messing it with irons, because it always takes the tint of the iron and causes a dark cast. When it is cooked and clear, see if it has enough manganese or if it is greenish. Once again, be warned that in order to make the crystal beautiful, you must always use manganese of Piedmont the way it is made for Murano, because the manganese of Tuscany and Liguria has more rust, which always make the melt dark. Therefore, always use manganese of Piedmont, since it is the best that is currently known in the art of glassmaking.

Now of manganese, you should only add a little bit and with discretion because too much would make the crystal a purplish wine color, and then would darken it and remove its splendor. Mix in the manganese to clean the glass until it becomes clear and glistening. It is a property of manganese when it is first added, to turn the melt green, this then forms the crystal and makes it appear a splendid white, because that greenishness removes the scum. However, you must add it little by little, in order not to spoil the crystal.

12　　　　LIBRO PRIMO.

ſi ritorna in padella pulita, & queſto tragiettare in acqua ſi deue fare più uolte ſecondo il biſogno, acciò il criſtallo ſia ſeparato da tutto il ſale, che queſto ſi rimette alla pratica del conciatore di fornacie: poi ſi laſci quocere per quattro, o ſei giorni, & il manco, che ſia poſſibile ſi meſti con ferro, perche ſempre piglia la tintura di eſſo, & lo fa tirare al nero, come è cotto, & chiaro guarda ſe ha aſſai manganeſe, & ſe verdeggiaſſe dagli del manganeſe.

Auuertendo, come s'è detto, che per fare il criſtallo bello, ſi deue ſempre pigliare manganeſe del Piemonte, come ſi fà a Murano, perche il manganeſe di Toſcana, & di Liguria hanno più del ferrigno, & fanno ſempre nero: & però ſi vſi ſempre manganeſe del Piemonte, che è il meglio, che hoggi ſia in notitia nell'Arte Vetraria: però del manganeſe ſe ne dia poco, & con diſcretione: perche farebbe il criſtallo in colore di auuinato, che poi tende al nero, & gli leua il ſua ſplendore, meſtiſi il manganeſe, & ſi laſſi pulire il vetro, tanto che venghi di colore chiaro, & lampante. Il proprio del manganeſe, quando è dato debitamente è di torne il verdeggiare, che fa ſempre il criſtallo, & gli fa apparire vn bianco riſplendente, perche gli toglie quella verdezza rozza; però di queſto ſe gliene dia a poco a poco; per non guaſtare il criſtallo, perche tutto queſto negotio conſiſte nella pratica de i valenti, & diligenti conciatori, che non ſi può dare, ne peſo certo, ne miſura, quando hauerai il criſtallo pulito limpido, & bello, faralo lauorare del continuo in vaſi, & lauori, come più ti piace: ma non con tanto fuoco, come ſi lauora il vetro comune, & ſopra tutto ſi operi, che il fuoco della fornace ſia chiaro, & ſenza fumo, & queſto di legne forte, & ben ſecche, & li ferri con che ſi lauora ſiene netti, & puliti, auuertendo non ritornare i colletti, doue è l'attaccatura delle canne, che ſempre vi rimane del ferro, nelle padelle del criſtallo, che lo faria diuentare nero : però ſi auuerta in particulare : anzi queſto vetro, oue è l'attaccatura del ferro delle canne ſi può mettere nel vetro comune, del quale ſi fanno lauori dozzinali, che in queſto poco importa. Queſto è il modo, di fare, & lauorare il criſtallo, come ho praticato ſempre mai.

A fare

This entire exercise depends on the practice of being a worthy and diligent glass compositioner, because neither sure weight nor measure can be given.

When you obtain a limpid and beautifully clean crystal work it continually to make vessels and items as most appeals to you, but not with as great a fire as when working common glass. Above all, ensure that the fire in the furnace is clear and without smoke, the firewood is hard and very dry, and the irons for working the crystal are neat and cleaned. Make sure never to return the neck, where the rod attaches to the glass, into the crucible of crystal, because there are always remains of the iron that will cause it to become dark; you must avoid this particularly. Indeed, put this glass attached to the iron rod into the common glass, and use it for mass-produced jobs of little importance. This is the way, to make and to work crystal, as I have practiced without fail.

A fare il Criſtallino, & vetro bianco, detto altrimenti vetro comune. Cap. X.

SE metterai in padella la fritta fatta di poluerino, hauerai vetro bianco, & bello, detto vetro comune, il modo di fare le fritte di Poluerino, e di Rocchetta ſi è detto chiaramente al ſuo luogo. Se metterrài la fritta di Rochetta, all'hora ſi darà vn vetro belliſsimo detto Criſtallo, quale e mezzo tra il vetro ordinario, & il bollito, altrimenti detto criſtallo : la fornace ſi ſtizzi ſempre con legne forte, & ſecche, guardandoſi dal fumo, che ſempre nuoce, & fa nero, & tanto al vetro bianco detto comune, quanto al criſtallino ſi dia la ſua dolſi, e parte di manganeſe del piemonte, & queſto preparato, come ſi è detto nel bollito chiamato criſtallo, acciò venga bello, perche il manganeſe gli toglie il verdiccio, & pauoniccio, che per ancora hanno ſempre mai queſti vetri, & il criſtallino almeno per vna volta ſi deue tragietare in acqua, che di coſi hauerai coſa bella, & chiara, l'iſteſſo ſe potrai fa ancora al vetro comune per hauerlo in perfetione, poi ſi ritornano in padella al ſolito, e come ſono puliti, e belli ſi faccino lauorare in quei lauori, che fa di biſogno, auuertendo, che il tragietare in acqua ſi fa ſecondo i guſti, che ſe bene ſi può fare ſenza eſſo, tuttauia, quando ſi deſidera vetri belli più dell'ordinario, queſto è neceſſario : perche oltre al farlo più bianco, lo calcina, che puliſce bene, & ha poche pulighe. Auuertendo ſopra ogni coſa, che ſe al criſtallino, & anco al vetro comune ſi darà a ciaſcuno da per ſe ſu la fritta a ragione di dieci libbre per cento di ſale di tartaro purificato, fa il vetro, & criſtallino ſenza comparatione più bello, che l'ordinario, e più tenero a lauorare, guardando nondimeno, che nel criſtallino non ſi buttino dentro i colletti, oue è l'attaccatura del ferro della canna, perche queſti ſempre fanno nero, ma ſi mettino nel vetro comune. Il ſale di tartaro ſi auuerta, di darlo quando ſi fa la fritta, & all'hora meſcolarlo con il tarſo, ò rena inſieme con il Poluerino, o vero Rochetta ſtacciata, & ſi faccia fritta al ſolito. Il modo di fare il ſale di tartaro purificato per l'opera ſopradetta è l'infraſcritto, cioè.

A fare

produced jobs of little importance. This is the way, to make and to work crystal, as I have practiced without fail.

~ 10 ~
To Make Crystallino and White Glass, Also Known as Common Glass

If you put frit made from polverino into a crucible, you will have a fine white glass called common glass. The way to make frit of polverino, and rocchetta I describe clearly in previous chapters. If you use the frit of rocchetta, this will make the finest glass, which is half-way between ordinary glass and bollito, also known as crystal. Stoke the furnace from time to time with hard and dry firewood. Be on guard to prevent it from smoking, which will always damage as well as darken common white glass.

As for the crystallino, a portion of manganese of Piedmont will make it smooth. Prepare this as described in the recipe for bollito crystal. Manganese also makes it beautiful because it removes the greenish and blue-greenish tint, that never the less these glasses and crystallino invariably have. If you throw it into water at least one time, what you will have will be beautiful and clear. The same is true for common glass, which once brought to perfection you should return to the crucibles for use. It will be bright, fine, and quite satisfactory to work in those jobs that require it. Take note that you should throw the glass into water after the assay. It can well be made without doing this, however, when a more than ordinary fine glass is desired it is necessary. Beyond becoming very white, it calcines and clarifies nicely with few impurities.

Above all else, take note that if you add 10 pounds per 100 of purified tartar salt to crystallino or to common glass frit, doing each type separately, it makes the crystallino and glass better beyond comparison and more pliant to work than the ordinary. Be vigilant never to throw the necks of the glasses into the crystallino; put them in the common glass instead as they always cause darkening. The neck is where the iron rod attaches to a piece for working. Also, take note to add the tartar salt at the point that you make the frit. Mix it in with the tarso or sand and the polverino or well sifted rocchetta salt and then make the frit as usual. The way to make purified tartar salt for this work is as follows.

~ 11 ~
To make Purified Tartar Salt

You should obtain tartar, which is also called gruma, from barrels of red wine in which it forms large lumps, however do not use powder. Roast it in earthenware pots amongst hot coals until it becomes calcined black and all its sliminess is roasted away. It then will begin to whiten, but do not let it become white, because if you do the salt will be no good. Calcine tartar this way and put it in large earthenware pans full of hot common water, or better yet in glazed earthenware pans and made to boil on a slow fire. You should do it in such a way that in 2 hours the level of the water will slowly decrease to one-quarter, at this point lift it from the fire, and leave it to cool and to clarify.

14   DELL'ARTE VETRARIA.

A fare il fale di Tartaro purificato. Cap. XI.

HABBISI del Tartaro, che altrimenti fi dice Gruma di botte di vino roffo, & fia gruma groffa, & non in poluere, quefta fi abbruci in pignatte di terra in fra carboni accefi, che venga calcinata nera, & fi abbruci ogni fua ontuofità, & che voglia cominciare a imbiancare, però non fia bianco, perche fe imbiancafsi non faria buono il fale, detto Tartaro cofi calcinato fi metta in catinelle di terra grande, piene di acqua comune calda, anzi in pignatte di terra vetriata, & fi faccino bollire a fuoco lento,& piano di tal maniera, che in due hore cali la quarta parte dell'acqua, all'hora fi leuino da fuoco, fi lafcino freddare, & chiarire l'acqua, laquale fi decanti, che farà vna lifcia forte, & fi ritorni nuoua acqua comune in dette pignatte, nel modo detto, & fopra le refidenze del tartaro, & fi bolla come fopra, & quefto fi rieteri, fino l'acqua non venga più falata, & carica di fale, all'hora quefte acque pregne di fale fi feltrino, & la rannata chiara, & feltrata fi metti in orinali di vetro a fuaporire in cenere di fornello a fuoco lento, che in fondo rimarrà vn fale bianco, quefto fale di nuouo fi folua in acqua comune calda, & fi lafci in catinelle a pofare per dui giorni, poi fi feltri, e fi ritorni di nuouo in orinali a fuaporire a fuoco lento, che in fondo rimarrà vn fale molto più bianco della prima volta, qual fale di nuouo fi folua in acqua comune calda fi lafci pofare per doi giorni, & poi fi feltri, & fi fuapori, come fopra in tutto, e per tutto, & quefto modo di foluere, feltrare, e fuaporare quefto fale di tartaro fi reiteri per quattro uolte, che all'hora farà vn fale bianchifsimo più della neue, & purificato in gran parte della fua terreftrieità, quale fale mefcolato con il Poluerino, o Rochetta ftacciata con la fua dofi di tarfo, o rena, farà la fritta, che in padella farà criftallino, & vetro comune molto più bello affai , che non fi fa fenza l'accompagnatura di quefto fale di tartaro, che fe bene fenza effo fi fa criftallino bello, tuttauia con quefto farà molto più bello.

A pre-

Now decant off the liquid, which will be strong lye and refill the pans containing the remains of the tartar with new common water. In the way stated above, boil as before and repeat the procedure until saltiness no longer charges the water. At this point, the [decanted] water is impregnated with all the salt. Filter the lye clear and put it in glass chamber pots to evaporate in the ash of the stove over a slow fire. In the bottom, white salt will remain. Dissolve this salt in new hot common water and leave it in the pans, letting it settle for 2 days. Then filter it and return it to chamber pots to evaporate over a slow fire. In the bottom, a much whiter salt will be left than the previous time. Now dissolve this salt in fresh hot common water and leave it to settle for 2 days. Evaporate, filter, and repeat everything as before. All in all, repeat this procedure four times to dissolve, filter, and evaporate the salt of tartar. This will make the salt whiter than snow and purified from the vast majority of its sediment.

When mixed with sifted polverino, or rocchetta, with its doses of tarso or sand, this salt will make a frit that in crucibles will produce the most beautiful crystallino and common glass, which one cannot make without the accompaniment of tartar salt. Without it [tartar], good fine crystallino can be made, nevertheless with it, it will be the absolute most beautiful.

~ 12 ~
To Prepare Zaffer, Which Serves for Many Colors in the Art of Glassmaking

You should get zaffer [Saxon blue] in large pieces and put it in earthenware oven-pans holding it in the furnace chamber for half a day. Then put it into iron ladles to inflame it in the furnace. Heat it well, then take and sprinkle it with strong vinegar. When cold, grind it finely over a porphyry stone into glazed earthen pots with hot water. Then wash more water over it always leaving the zaffer to settle in the bottom.

DELL'ARTE VETRARIA. 15

A preparare la Zaffera, che ferue per più colori nell'Arte vetraria.   Cap.   XII.

PIGLISI lá Zaffera in pezzi grofsi, & mettafi in tegami di terra tenendola nella camera della fornace per vno mezo giorno, di poi fi metta in vna cazza di ferro a infiammare nella fornace, & fi caui, & cofi calda fi sbuffi con aceto forte, poi come è fredda fi macini fottilmente fopra porfido, & in catinelle di terra inuetriata con acqua calda fi laui, & a più acque, lafciando fempre pofare la Zaffera in fondo, poi fi dècanti pianamente, che cofi portera via la terreftreita, & immonditie della Zaffera, & la parte buona, & tintura della Zaffera rimarrà in fondo, laqualè cofi preparata, & purificata tingnera affai meglio, che prima, facendo tintura limpida, & chiara, quefta Zaffera fi afciughi, & fi ferbi in vafi ferrati al fuo vfo, che farà affai meglio che prima.

A preparare il Manganefe per colorire i vetri. Cap.   XIII.

HABBISI Manganefe del Piemonte, che quefto è il meglio, di tutti li Manganefe, che oggi fieno in notitia nell'arte Vetraria: che in Venetia fe ne troua fempre copia: poi che a Murano non fi vfa altro Manganefe. In Tofcana, & in Liguria ne fa affai, ma tiene molto di ferro, e fa nero, & brutto : & quel del Piemonte fa vno auuinato belli simo, & da vltimo lafcia il vetro candido, & gli toglie il verdegnolo, & azzurigno: adunque quefto Manganefe del Piemonte fi metta cofi in pezzi, come è in cazza di ferro, & fi faccia reuerberare nella fornace, & cofi infiammato fi sbuffi con aceto forte, poi fi macini fottilmente, come la Zaffera, & fi laui a più acque caldo come s'è detto della Zaffera, fi & poluerizi, & fi ferbi in vafi ferrati al fuo vfo, & bifogno.

A fare

Now gently decant, to carry away the sediment and impurities of the zaffer. The good part and pigment of the zaffer will remain in the bottom. The pigment remains are now prepared and purified far better than at first, which will make clear and limpid pigment. This zaffer should be dried and kept in sealed vessels for use, which will be much improved over the original.

~ 13 ~
To Prepare Manganese for Coloring Glass

You should have manganese from Piedmont; this manganese is the best of any that are currently known in the art of glass. In Venice, you can always find it in abundance, since on Murano they do not use any other manganese. In Tuscany and Liguria there is plenty, but it contains a lot of iron and causes a foul darkening, while that of Piedmont makes a beautiful purple wine color, and in the end it leaves the glass a pure white removing any greenish or bluishness. Therefore, this manganese of Piedmont is broken into pieces, put in iron ladles, and made to break down in the furnace. Once it has been well inflamed, sprinkle it with strong vinegar, then grind it like the zaffer and wash it with more hot water as was described for the zaffer. Pulverize it and store in sealed vessels for future use.

~ 14 ~
To Make Spanish Ferretto, Which Serves to Color Glass

To make ferretto requires nothing more than a simple calcination of copper, where the metal is opened so it can impart its pigment to the glass. When this calcination is done well there is no doubt to anyone that a very interesting diversity of colors appear in the glass. This calcination may be done many ways, but I will put down two that are not only easy, but have been used by me many times for a variety of beautiful effects in the glass. The first method follows here.

You should get thin shims of copper the size of a florin [coin] and have one or more goldsmith's crucibles ready. In the bottom of them, you should make a layer of pulverized sulfur, then a layer of the shims, and over them another layer of pulverized sulfur, and one of copper shims, and so on. With this method, fill the crucible in what is otherwise called a stratification. Cover the crucibles, coat them well with refractory clay, dry, and place in an open oven to vent amongst burning coals and put a strong fire to them for 2 hours. Allow them to cool and you will find calcined copper, which will come apart with the fingers, as if it were dry earth. It will rise in color to a blackish-reddishness; this copper should be ground fine, passed through a sieve, and kept well secure for use in coloring glass.

~ 15 ~
Another Way to Make Spanish Ferretto

This second way to make ferretto is much more laborious than the first; nonetheless, it will produce extraordinary effects in the glass. Instead of stratifying the copper in the crucible with sulfur, stratify it with vitriol [sulfur anhydride]. Now calcine it,

16   LIBRO PRIMO

A fare il Ferretto di Spagna, che serue ne i colori de i vetri.
Cap.   XIIII.

FARE il Ferretto non è altro, che vna semplice calcinatione di rame a effetto, che il metallo aperto possa communicare nel vetro la sua tintura, qual calcinazione, quando è ben fatta non è dubbio alcuno, che nel vetro fa apparire colori diuersi, & molto vistosi : tal calcinazione si fa in più modi, però io ne metterò duoi, non solo facili, ma per me vsati molte volte, con effetti assai belli nel vetro, de i i quali il primo è l'infrascritto, cioè.
Habbisi lamine di rame sottili della grandezza di vna piastra Fiorentina, & habbisi vno, o più coreggioli da orefici, & nel fondo di essi coreggioli farai vn suolo di zolfo poluerizato, poi vn suolo di dette lamine, & sopra vn'altro suolo di zolfo poluerizato, & vno di lamine di rame, come sopra. & con questo ordine empi il coreggiolo, che altrimenti si dice stratificare, questo coreggiolo per sopra si cuopra, & si luti bene, & asciutto si metta in fornello aperto a vento in fra carboni ardenti, & se li dia il fuoco gagliardo, per due hore, si lasci freddare, & trouerrai il rame calcinato, & si spezzera con le dita, come se fusse di terra secca, & farà gonfiato in colore nericcio, & rossigno, questo rame si macina minuto, & si passa per staccio fitto, & si serba ben custodito a'bisogni de colori de vetri.

Altre modo da fare il detto Ferretto.   Cap. XV.

QVESTO secondo modo di fare il ferretto, se bene è più laborioso del primo, tuttauia fara il suo effetto nel vetro più che ordinario.
Adunque il rame in cambio di stratificarlo con zolfo nel coreggiolo si stratifichi con Vitriolo, & poi si calcini lasciando stare nella camera della fornace vicino all'occhio per tre giorni, poi si caui, & si ritorni a stratificare pure con nuouo vitriolo, & si tenga a reuerberare, come sopra, & questa calcinatione con vitriolo si reiteri sei volte, che all'ora, s'hauerà vn ferretto nobilissimo, che ne colori farà effetti più che ordinarij.

A fare

leaving it in the furnace [annealing] chamber, near the 'eye' [heat-vent] for 3 days. Remove it, stratify it again with new sulfur anhydride, and leave it to break down as above. Repeat this calcination with the anhydride six times at which point you will have a most noble ferretto whose colors will produce extraordinary effects.

~ 16 ~
To Make Iron Crocus, Also Known as Crocus Martis, to Color Glass

Crocus martis is nothing other that a refinement and calcination of iron. A means by which its pigment, which in glass is a deep ruby red, is opened and imparted to the glass. It not only manifests itself but makes all the other metallic colors as well, which ordinarily hide and are dead in the glass, dance in resplendent apparition. Since this is the way to make the hidden metallic colors appear, I have put down four ways to make it, and the first is:

Obtain some iron filings, although steel is better if you can get it. Mix this well with three parts pulverized sulfur and put into crucibles as described above for the ferretto. Keep it in the oven to calcine and to consume all of the sulfur thoroughly, which will happen quickly. Leave it between burning coals for 4 hours, then remove, pulverize and sift it through a fine sieve at which point you should put it into

DELL'ARTE VETRARIA. 17

A fare il Croco di ferro, altrimenti detto di Marte, per i colori del vetro. Cap. 16.

IL Croco di Marte, non è altro, che vna futtigliazione, & calcinazione di ferro : per mezzo della quale la fua tintura, che in vetro è rubicondifsima fi apra di maniera, che comunicatafi con il vetro, non folo manifefti fe ftefla ma faccia, che tutti gli altri colori metallici, che per ordinario nel vetro fariano occulti, & morti, apparifchino vaghi, & rifplendenti: perche quefto è il mezo di fare apparire l'occultezza metallica: io metto quattro modi di farlo, & il primo è

Habbi limatura di ferro, potendo hauere di acciaio è meglo, quefta fi mefcoli bene con tre parte di zolfo poluerizzato, & in coreggiuolo, come fopra fi è detto del ferretto fi tenga in fornello a calcinare, & abbruciare tutto il zolfo benifsimo, che fuccede prefto, & fi lafci ftare in fra carboni ardenti per quattro hore: poi fi caui, & fi poluerizi, ftacci per ftaccio fitto all'hora fi metti in coreggiuolo per fopra coperto, & lutato, & fi tenga nel era della fornacie preffo al occhio, o vero lumella per quindici giorni, o più, che all'hora piglia vn colore roffigno pauonazziccio quafi purpureo quefto ferua in vafo ferrato per vfo de i colori de i vetri, perche fa molti belli effetti.

A fare Croco di Marte in altra maniera. Cap. 17.

QVESTO fecondo modo di fare il Croco di Marte con tanta facilità non fi deue difprezzare, anzi ftimare affai, poi che il Croco fatto in quefta maniera nel vetro fa apparire il vero rubicondo di fangue, & il modo di farlo è quefto.

Habbi limatura di ferro, potendo hauere di acciaio è meglio, quefta fi mefcoli bene in tegame di terra, con aceto forte, cioè fi irrori folamente tanto, che fia inhumidita per tutto, poi diftefa in detto tegame, fi tenga al fole, che fi afciughi, & non effendo il Sole fcoperto, fi lafci cofi all'aria, che come è fecca, all'hora fi torni a peftare, che farà alquanto ammaffata, & con nuouo aceto fi irrori, &
C   inhumi-

covered crucibles, coated with clay and keep them in the furnace [annealing chamber] near the 'eye' for 15 days or more. At this point it takes on a reddish-peacock like color, almost purple. Store it in a closed vessel for use in coloring glass, since it produces so many beautiful effects.

~ 17 ~
To Make Crocus Martis in Another Manner

This second way to make crocus martis although very easy should not be disparaged, but rather highly regarded, since crocus produced this way causes the glass to appear a rather bright blood red. The way to make it is like this:

Obtain some iron filings, although steel is better if you are able to get it. Mix it well in an earthenware oven-pan with strong vinegar, which is sprinkled only until it is moist throughout. Now spread it out in the pans, and put it in the sun to bake. When the sun is clouded, leave it in the open air to dry. Now turn it into powder. If it leaves any hard lumps, sprinkle and moisten them with new vinegar, leave it to dry again and then pulverize, as above. This work should be repeated eight times, then grind and sift it through a fine sieve, which will make a very fine powder the color of brick dust. Store this in well-sealed vessels for use in the coloring of glass.

~ 18 ~
Another Way to Make Crocus Martis

This third way to make crocus martis, with aqua fortis [nitric acid], is a way in which the profound color of the iron is manifest perhaps more than seems credible. The true proof of this must be seen and experienced in the glass itself.

Therefore, the filings of iron or steel are put in glazed earthenware pans, wet with nitric acid, and left in the sun to dry. Now turn them to powder, sprinkle with nitric acid again, and let dry. Repeat this procedure several more times, and it will become deep reddish as was described for the crocus made with sulfur. Now pulverize and sift it, and store for future glass coloring needs.

~ 19 ~
To Make Crocus Martis in Another Manner

This is the fourth and last way I show how to make crocus martis, and perhaps the best of all.

18   LIBRO PRIMO
in humidifca, & fi torni à feccare, & poluerizzare, come
fopra, quefta opera fi reiteri per otto volte, poi fi macini,
& paffi per ftaccio fitto, che farà vna poluere fottilifsima
in colore di matton pefto, quefta fi ferbi in vafo ben ferrato per vfo de i colori de i vetri.

Altro modo di fare Croco di Marte.   Cap. 18.

QVESTO terzo modo di fare il Croco di Marte con
acqua forte, e modo, per ilquale il profondo colore del
ferro fi manifefta più di quello non par forfe credibile, &
nel vetro fe ne vede la vera efperienza, & proua.
Adunque la limatura di ferro, o acciaio in tegame di terra
inuetríato fi irrori con acqua forte, & fi tenga al Sole a
feccare, fi torni a poluerizzare, & a sbruffare con acqua forte, & fi afciughi, & fi reiteri cofi più volte, poi fi rubifichi
come fi è detto nel Croco fatto con il zolfo, poi fi poluerizzi, & ftacci, & fi ferbi per il bifogno di colorire i vetri.

A fare croco di Marte in altra maniera.   Cap. 19.

QVESTO è il quarto, & vltimo modo, che io moftro
per fare il Croco di Marte, & per auuentura il meglio
di tutti, però ciafcuno delli modi per me moftrati in fua
operatione non folo è buono, & perfetto, ma neceffario
ancora per i diuerfi colori, che fono neceffari farfi quoti-
dianamente nel vetro, & per far quefto, foluafi adunque
in acqua forte, fatta regis con fale armoniaco al folito, come fi dirà nelle regole del calcidonio, limatura di ferro, o
vero acciaio, che è meglio, in vafo di vetro ben ferrato
fi tenga per tre giorni, & ogni giorno fi agiti bene: però fi auuerta quando fi mette la limatura fopra dett'acqua
di fare pianamente, perche gonfia affai, & porteria pericolo di far crepare il vetro, o vero di vomitare tutta fuo-
ra, però fi vadia cauto nel metterla: in capo di tre giorni
fi fuapori l'acqua a fuoco lento, che nel fondo farà vn croco di Marte nobilifsimo per le tinture di vetri ftupendi-
mente, quale fi ferbi per fuo vfo.

A cal-

However, every one of the ways shown by me in this book is not only good and perfect, but also necessary for the various colors that are required to be made daily in glass.

To make this you should therefore dissolve filings of iron, or steel which is even better, in nitric acid made into aqua regis with ammoniac salt [ammonium chloride] as per usual- which will be described in the recipe for chalcedony. Keep it for 3 days in a well-sealed glass vessel, and every day agitate it well. But be careful and put the filings into the acid slowly, because it will swell quite a bit and could violently erupt or put the glass in danger of cracking, so exercise caution in adding it. At the end of 3 days evaporate the water over a slow fire, and in the bottom, it will leave the noblest crocus martis for pigments of the most wonderful glasses, which you should then store for use.

~ 20 ~
To Calcine an Orpiment Called Tinsel, Which in Glass Makes the Celestial Color of the Blue Magpie

As is very well know, the orpiment known as tinsel is composed of copper, which from calamine [zinc silicate] takes a tint in color similar to gold. The calamine not only tints the copper, but incorporating it adds quite a bit of weight. This augmentation, when it is well calcined, gives glass a color that is quite delightful to see, holding the middle between aquamarine and the color of the sky when it is very clear and serene; a thing of beauty. You must be very diligent in calcining it, so here is the way to make it step by step.

Take orpiment, also known as tinsel, and to save money purchase some that has already been used for decorative wreaths and garland. Cut this into small pieces with scissors, and then put it into covered chamber pots coated with clay, amongst the coals in a strong fire. I put it in the burning coals of the furnace on the side where it is stirred up, and I leave it be for 4 days in a large fire, but not hot enough to fuse, because if it melts, all the work will be lost. After the prescribed time it will be calcined quite well.

Next, I grind it into powder and pass it through a fine sieve, then I mill it over a very fine porphyry stone, at which point it becomes a black powder. I then spread it in oven pans and hold it in the annealing chamber close to the 'eye' for 4 days. Remove the ash that accumulates on top, pulverize, sift, and store it for use. The test for good calcination is that when sprinkled into the glass melt it causes [the glass] to swell quite a bit. If it does not make the glass swell and boil vigorously, this is sign of poor calcination or too much burning, in either of these cases, if it does not make the glass boil it will not color well. Therefore, this is a warning for all to heed in the practice.

20   LIBRO PRIMO

A calcinare il medefimo Canterello in altra maniera, per fare il roßo tranfparente, il giallo, & il Calcidonio.
Cap. 21.

PIGLIA il Cantarello fopradetto tagliato con le forbice minutamente, & in correggiuolo fi ftratifichi con zolfo poluerizzato, fi metta in fra carboni accefi, io lo metteuo nel tizonaio della fornace à calcinare per ventiquattro hore: poi lo peftauo, & ftacciauo, & lo metteuo in tegame di terra coperto nell'era della fornace per dieci giorni a riuerberare vicino al occhio, poi lo poluerauo, & macinauo, & cofi febauo al fuo bifogno.

Acqua Marina in vetro; colore principale nell'arte.
Cap. 22.

L'ACQVA marina, anzi il colore detto acqua marina è vno de' principali colori, che fi dia al vetro, & à volere, che fia bello, & di tutta proua fi deue fempre mai fare nel bollito, altrimenti detto criftallo artifiziale, perche nel vetro comune non vien bello, & nel criftallino fe bene vien più bello, che nel vetro comune, tuttauia folo nel bollito detto criftallo viene in perfetione. Auuertendo, che ouunque fi vuole fare l'acqua marina, per niuna maniera a tal vetro fe li dia dato il manganefe per prima: perche quefto dato etiam che il fuoco poi l'habbi confumato, tuttauia lafcia vna qualità nel vetro, che fempre fa negreggiare l'acqua marina, & gli da imperfetione, & bruttezza grande. Adunque per fare vn'acqua marina di garbo, & bella fi piglierà la fritta di criftallo, & fi metterà in padellotto, fenza dar gli punto di manganefe, & come è fufo, & pulito bene per fopra fa fale, quale nota fopr'il vetro in forma d'olio, quefto fi caui con le cazze di ferro da i conciatori, come ben fanno, & fi caui tutto con ogni diligenza perche re ftandouene faria il colore brutto, & vntuofo, e fi lafci, che il vetro pulifca bene in tutta perfetione, all'hora per efempio a un padellotto di libbre venti di criftallo in circa fi pigli oncie fei di canterello calcinato, & preparato,
come

~ 21 ~
To Calcine the Same Foils Another Way in Order to Make Transparent Red, Yellow, and Chalcedony

Take some foil and as above cut it with the scissors into tiny pieces and put it in amongst burning coals. I put it in the glowing coals of the furnace to calcine for 24 hours. Then I grind it and sift it, and put it in covered earthenware oven-pans in the annealing chamber of the furnace close to the 'eye', for 10 days to break down, and then I grind and mill it and finally store it for use.

~ 22 ~
Aquamarine in Glass, a Principal Color in the Art

Aquamarine, or rather the color called aquamarine is one of the principal colors used in glass, and to ensure that it will always come out nicely you must always make it with bollito, which is also called artificial crystal. Since in common glass it does not come out nicely, and while it comes out much better in crystallino than in common glass, it never the less only comes out perfectly in bollito crystal.

Take care that whenever you want to make aquamarine, that no one has first added manganese to the glass, because once it has been added even though the fire will then consume it, it never the less leaves a quality in the glass that always makes the aquamarine darken and will cause large ugly inclusions. Therefore in order to make a nice handsome aquamarine, take crystal frit and put it in a large crucible, without giving it a spec of manganese, and allow it to fuse and clarify well. It will form a salt on top, which you will note on the surface of the glass in the form of an oil. Take this off with the kind of iron ladle that the glass compositioners know and use so well. Remove it completely with the utmost diligence because any residue you leave will produce an ugly greasy color.

Let the glass clarify to full perfection. Now for example, add 20 pounds of crystal to a large crucible with approximately 6 ounces of calcined tinsel, prepared as was described previously in chapter 20 for making a celestial or blue magpie color. To this calcined tinsel, add one-fourth part of prepared zaffer. Mix these two powders well and add them to the crystal in three portions, because the powder causes the glass to swell quite a bit. Then you must stir the glass with a paddle, but when the tinsel is calcined well and as directed, it swells so much that it could make all the glass go out of a large crucible. So use diligence in this. Wait for the glass to settle down and let the color incorporate for 3 hours.

Then turn and blend the glass, and take some of it out. Test it to see whether it is sufficiently charged, and whether to make the color more or less intense. Because in order to make vessels and drinking glasses where the glass is thin, you must really load it with a lot of color, but for making large cane for beads not so great a charge of color is necessary. For making thin cane for small beads, you must charge it well with color. In working the glass, you must apportion it with more or less color according to the purpose it must serve. This should be left to the judgment of whoever will be doing the work. Because it always settles, do a test and come down on the side of too little because you can always add more color to the glass after it has clarified. 24 hours after the color is added, you must work it. Be vigilant; foremost is the task of stirring the glass well in the bottom of the crucible, since the color will condense and vary throughout. When you allow the glass to stand still, the color settles to the bottom, and on top it becomes color free. Therefore, from time to time stir it as necessary. Use the same rules, instructions, and doses to color the large crucibles of 300 or 400 pounds of crystal for the furnace when producing beadmaking cane.

I demonstrated this method of making aquamarine in Florence in the year 1602, at the Casino, and I made many batches of it for beadmaking cane, which always resulted in a most beautiful color. Take note, than in Murano for beadmaking cane they take half crystal frit and half rocchetta frit, and nevertheless still get a nice aquamarine; but in pure crystal it is the most beautiful.

~ 23 ~
A Celestial, or Rather Blue Magpie Color

You should have a large crucible in the furnace, or rather a crucible of clarified glass made from frit of rocchetta or Spanish barilla, the Levantine rocchetta is best. Once the glass has clarified very well, to a large crucible of 20 pounds [of glass] take 6 ounces of tinsel, calcined as was just described in chapter 20 (how to make a celestial or blue magpie color) and add it to the glass as was prescribed for the aquamarine. Follow the instructions, and take note that you must de-salt this glass, so skim it with great care, as is well known to furnace compositioners.

22   LIBRO PRIMO

Colore celeste, o vero di gazzera marina.   Cap. XXIII.

HAbbisi in fornace un padellotto, o uero padella di vetro pulito di fritta di rocchetta o soda di spagna, la rocchetta di Leuante fa meglio, come il uetro è pulito benissimo, all'ora per esempio a un padellotto di libbre uenti pigli once sei di tremolante calcinato per sè solo come s'è detto al capitolo 20. per fare il colore celeste, o uero di gazzera marina, & si dia al uetro, come s'è detto nell'acqua marina in tutto, e per tutto, auuertendo, che questo uetro sia desalato con la cazza con ogni diligenza, come ben sanno i conciatori di fornace, all'ora si fa un colore celeste, o uero di gazzera marina marauiglioso, il quale colore si fa più carico, e men carico, secondo i lauori a che deue serui re, come è ben noto a quelli dell'arte, in capo di due hore si torna a mescolare il uetro bene, e sene caui una proua, e si guardi s'è carico tanto, che basti, per poterlo caricare con nuoua poluere, o non caricare, come sta a satisfatione si lasci così per 24. hore, poi si mescoli, & si lauori, che questo farà un colore celeste, o uero di gazzera marina bellissimo & uariato da gl'altri colori, che si fanno nell'arte uetraria. questo colore ne tinsi moltissime padelle in Pisa l'anno 1602. & sempre uenne colore bello, & di tutta proua.

Ramina rossa, che serue a più colori in vetro.   Cap. 24.

SI pigli rame in piastre sottile, & si metta nelli archi della fornace, & ui si muri dentro, & si lasci tanto, che detto rame si calcini bene per se solo, con il semplice fuoco, però che non fonda, ne habbia fuoco di fusione, che in tal caso non si farebbe cosa buona come è calcinato si pesti, e poluerizzi, che uerrà in poluere rossa, la quale si serbi alli suoi usi che nell'arte uetraria sono molti, & tutti necessarij.

Ramina di tre cotte per i colori di vetro.   Cap. 25.

LA sopradetta ramina rossa si metta infornello, o uero nel era della fornace presso all'occhio in tegoli di terra cotta, o uero tegami di terra cotta, si lasci a calcinare per quattro
giorni

At this point, it will make a marvelous celestial or blue magpie color, which you should make more or less tinted with color according to the purpose that it must serve, as is also very well known to those in the art. After 2 hours turn and blend the glass well, and take some out to do a test. See if the dosage is sufficient, and whether you should charge it with new powder or not, so it is to your satisfaction. Then leave it, and after 24 hours stir, and your work will be a celestial color, or more properly blue magpie; a color as beautiful and varied as any in the art of glassmaking. With this color, I tinted quite a few crucibles in Pisa in the year 1602. It always proved to be a very nice color.

~ 24 ~
A Red Copper Scale that Makes Many Colors in Glass

Take small pieces of copper and put them inside the arches of the furnace. In that place, they will be within the walls. Leave them that way until each piece of copper is well calcined, using a simple fire. Do not use a fusing fire, or one that could melt them, in which case they would not come out well. When they are calcined, grind and pulverize them and they will become a red powder. Store it for the many necessary uses it has in the art of glassmaking.

~ 25 ~
Thrice Baked Copper Scale for Glass Colors

Put the above mentioned red copper scale into an oven, or better still in the annealing chamber of the furnace near the 'eye', on terracotta tiles or even on terracotta oven-pans. Leave it to calcine continuously for 4 days; it will become a black powder that lumps together. This must be freshly ground and sifted with a fine sieve and returned to calcine as before in the annealing chamber of the furnace, leaving it 4 or 5 days. At this point, it will not stick together any more and it will no longer be black but grey, and turn to powder. This is called 'thrice baked' copper scale with which are made aquamarine, emerald green, the Arabic color called turquoise, which is a very serene sky color, and many others. However, you must take care in the third calcination that it is not done too much, or too little, because in either case it will not color the glass well. The sign of perfect calcination is to sprinkle some of it over crucibles of clarified glass, even in larger crucibles; it will swell and boil suddenly. If it does not give this sign then it is no good, nor well calcined, so you must be sure that it shows this sign and then it will be perfect.

DELL'ARTE VETRARIA.   23

giorni continui, che uerra in poluere nera, & attaccata in ſieme, queſta ſi peſti di nuouo, & ſi ſtacci con ſtaccio fitto, & ſi ritorni a calcinare come ſopra nel'era della fornace laſciandola quattro cinque giorni, che all'hora non ſi attacca piu inſieme, & non è tanto nera, ma bigiccia, & ſi ſpoluera da ſe medeſima queſta ſi dice ramina di tre cotte, con la quale ſi fa l'acqua marina, il uerde ſmeraldino, il colore arabico detto turchino, o uero aierino molto uiſtoſo, & molti altri colori, però ſi auuerta nella terza calcinazione, che non ſia troppo, ne poco calcinata: perche in tal caſo non coloriſce bene il uetro, & il ſegno, che ſià calcinata a perfezione, e che dàtane di eſſa ſopra il uetro pulito nelle padelle, o uero padellotti, lo fa gonſiare, & bollire ſubito, come non da queſto ſegno non è buona, ne ben calcinata, però ſi auuerta che uenga a queſto ſegno per hauerla in ſua perfezione.

Acqua marina in Criſtallo, Artifiziale altrimenti detto bollito Cap. 26.

AVNO padellotto di fritta di Criſtallo, che non habbia hauto manganeſe, & che ſia deſalato bene con ogni diligenza: perche io non vſauo mai tragettare il criſtallo quãdo vi uoleuo fare l'acqua marina: ma lo faceuo in quel cãbio deſalare con ogni diligenza, & ſia per eſempio di libre quaranta, quando è fuſo, & pulito beniſſimo, ſi habbi once dodici di ramina di tre cotte ſopradetta nel capitolo 25. & in eſſa ſi meſcoli meza oncia di zaffera preparata, come ſi moſtra al Cap. 12, ſi vniſchino bene inſieme queſte due polueri, & ſi dieno in quattro volte al padellotto, che coſì il vetro la piglia meglio, ſi meſcoli bene il vetro cõ ogni diligẽza, & ſi laſci ſtare per dua hore, & poi ſi torni a meſcolare di nuouo, & ſe ne caui una proua, & ſe il colore è carico à baſtanza, ſi laſci ſtare, che ſe bene l'acqua marina pare uerdeggi, tuttauia il ſale, che è nel uetro mangerà, & conſumerà detto uerdeggiare & ſempre farà tendere all'azzurro, & in capo di 24. hore ſi porta lauorare: per quello effetto, che ſarà fatta, & ſecondo il ſuo colore, più carico: e men carico, che l'haurai: perche i colori ſi caricano ſempre mai con dare nuoua poluere, & ſi ſcaricano con cauare del uetro tinto
to

~ 26 ~
Aquamarine in Artificial Crystal, Also Called Bollito

Use a large crucible of crystal frit that has never had manganese added but has been meticulously well de-salted, because I never use crystal that has ever been thrown [into water] when I intend to make aquamarine. Instead, I de-salt with great care.

As an example for 40 pounds, fuse and clarify it very well, then have 12 ounces of 'thrice baked' copper, described above in chapter 25, and with it mix ½ ounce of prepared zaffer, as shown in chapter 12. Unite these two powders thoroughly and add them to the crucible in four portions so that the glass will better accept them. Stir the glass well with full diligence, let it stand for 2 hours and then stir it well again. Do a test and if the color is sufficiently charged, leave it alone. Although the aquamarine will seem greenish, never the less the salt in the glass will eat and consume this greenishness and it will always incline to blue if you allow it to work on itself for a span of 24 hours before it is used.

It should have more or less charge of color in accordance to the way it will be fashioned. Since you can always strengthen colors by adding new powder and abate them by removing some of the tinted glass and replacing it with un-tinted white of the same nature and quality, you can restore any of the colors to their desired state. It is not possible to teach this with precise rule, but must rest with the discretion of the glassworker. I have practiced this method of making aquamarine many times without ever failing. Into the crystal frit mix one-half part of rocchetta frit to produce a beautiful aquamarine for beadmaking, but when pure bollito is used, it is the most marvelous and beautiful.

~ 27 ~
General Warnings for All the Colors

In order for colors to emerge in full beauty and perfection:

First, take note that every new crucible large or small, when heated for the first time leaves a residue of sediment in the glass. Any of the colors that are then made in them will be covered with grime and filth, but you can always glaze crucibles that are not too large with a coat of white glass, as is well known to the compositioners. By the second time the crucibles are used they will have lost that foul scum.

Second, take note, that these same crucibles large and small, should each serve a single color. You must never add glass in order to make another color. As an example, in any crucible than serves for yellow, you should never try to make garnet red, one that serves for garnet will make a very poor green for you, one that serves for red, will not make you a very good aquamarine, and so forth for all the colors. Therefore, every color must have its own crucible, small or large, and in this manner, the colors will become closer to perfection.

Third, be sure that you calcine powders properly, that is neither too much nor too little, because in either of these cases they will not color well.

Fourth, be sure to use the required proportions and doses, and make mixtures to correct proportion. The furnace should be hot and stoked with hard dry firewood, because the green and soft wood makes little heat and blows smoke over everything.

Fifth, be sure to divide the colorants. Use part in the frit and use part in the glass when it is fused and well clarified.

There are still other warnings that will be discussed in their own places, when their particular colors are dealt with.

24   LIBRO PRIMO

to, e mettene del bianco non tinto, & della medefima natura, & qualità, & così tutti i colori fi riducono al fegno
defiderato, che quefto non è pofsibile poterfi infegnare cõ
regola precifa, ma refta nella difcrezione di chi lauora.
Quefto di fare l'acqua marina è ftato fperimentato molte
volte da me fenza mai fallirmi, & fe con la fritta di criftallo fi mefcolerà la metà di fritta di rocchetta uerrà acqua ma
rina bella per conteria, però nel folo bollito è bellifsima'a
marauiglia.

Auuertimenti generali in tutti i colori. Cap. 27.

ACCIO i colori uenghino in tutta bellezza, & perfezione, fi auuerta, che ogni padellotto, o padella nuoua, che per la prima volta s'inforna, lafcia un rozume nel
uetro, della fua terreftreità, che tutti i colori, che ui fi fanno
pare fieno faluatichi, & rozzi, però i padellotti, che non
fono molto grandi fempre fi poffono inuetriare con uetro
bianco colato, come ben fanno i conciatori, però alla fe-
conda uolta le padelle hanno perfo quel faluatico, &
rozzume. Secondo fi auuerta, che quei padellotti, & padelle, che feruano à un colore, non ui fi deue mettere uetro, per fare un'altro colore: per efempio in padellotto, o
padella, che ferue al giallo, non è bene poi farui l'ingranato, uno che ferue all'ingranato non è bene farui il uerde:
uno, che ferue al roffo, non è bene farui l'acqua marina, &
così di tutti i colori: ma ciafcuno colore uorrebbe hauere
la fua padella, o padellotto, che in quefta guifa i colori uen
gono più perfetti. Terzo, che le polueri fieno calcinate à
ragione, cioè ne troppo, ne poco, che in qual fi fia di que-
fti cafi non tinghanno bene. Quarto, che fi dia la debita pro
portione, & dofi, e le mefcolanze fi faccino à proportione,
le fornaci uadino calde, & fieno ftizzate con legne fecche
& forti : perche le legne uerdi, e dolci con il fuo poco calo
re, & fumo guafta ogni cofa. Quinto auuertire, che una
parte de colori fi da in partita, cioè nella fritta, & una parte
fi da al uetro, quando è fufo, e ben pulito, ui fono altri au
uertimenti ancora, che fi diranno a' fuoi luoghi, quando fi
tratterà particolarmente de i colori.

A fare

~ 28 ~
To Make Thrice Baked Copper Scale with More Ease and Less Cost than Before

You should take copper scales, which are the flakes that the kettle-smiths make when they hammer pails, cups and other works of copper, that are hot-worked in the fire. This flake, which falls from these works when they are struck, is called copper scale. It costs much less than solid copper, of which the scale described previously is made. In order to calcine it is not necessary to open and close the arches of the furnace, as in the previous recipe, which causes quite a bit of trouble and disturbance to the furnace.

A fare Ramina di tre cotte con più facilità, & manco ſpeſa della ſopradetta . Cap. XXVIII.

PIGLISI Ramina, che è la ſcaglia, che fanno i calderai, quando battono ſecchie, mezzine, & altri lauori di rame, che rinfocolati i lauori gli battono, quella ſcaglia, che caſca ſi chiama ramina, laquale coſta manco aſſai del rame ſodo, del quale è fatta la ramina indietro deſcritta, & per calcinarla non occorre ſmurare, & rimurare li archi della fornace, come nella ſuddetta, coſa di molto incommodo, & diſturbo della fornace. Si pigli adunque queſta ramina, che ſia netta, & pulita da ogni terra, e ſporchezza, & ſia lauata con acqua calda più volte dalla ſua terreſtreità, & rimanghi la ramina netta da ogni immonditia, & all'hora ſi metta in tegoli di terra cotta, o tegami di terra cotta, ſi tenga nel'era preſſo al occhio, o vero in fornelli fatti a poſta. Io in Piſa haueuo fatto fare vno fornello piccolo a foggia di vna piccola calcara, oue calcinauo per volta venti, e venticinque libre di queſta ramina, & in poche hore. Però nella era preſſo la lumella della fornace vi ſi laſci ſtare per quattro giorni, poi ſi rinuoua, & ſi peſti beniſſimo, facendola paſſare per ſtaccio fitto, & di nuouo ſi ritorni in tegoli, o tegami di terra, come ſopra al medeſimo fuoco, & calore per quattro giorni, ché verrà in poluere nera, & ſi ammaſſerà inſieme, ſi peſti, & ſtacci per ſtaccio fitto, & in tegoli di nuouo ſi ritorni nel medeſimo luogo, & calore per quattro giorni, all'hora la ramina farà ottimamente preparata con manco faſtidio, & ſpeſa della ſopradetta, & fara nel colorire il medeſimo effetto in tutto, & per tutto, auuertendo per prima di calcinarla hauerla beniſſimo lauata da ogni terreſtreità, come ſi é detto: il ſegno quando è ben preparata farà, che faccia gonfiare il vetro, & bollire aſſai quando ſe li dà all'hora é ben preparata.

Acqua marina in Criſtallo bella con la ſopradetta Ramina. Cap. XXIX.

AVno padellotto di libbre ſeſſanta di fritta di criſtallo, che ſia deſalato beniſsimo, come ſopra s'è detto, & non
D   traggiet-

Therefore, use this copper scale, that you have cleaned free of all dirt and filth, and washed with hot water many times to remove any sediment. What remains are the clean copper scales free from all contamination. At this point, put them in terracotta oven pans or on terracotta tiles, in the annealing chamber near to the 'eye', or even better in ovens made specifically for the purpose. In Pisa, I contrived to make a little oven in the form of a small frit kiln, in which at times I calcined 20 to 25 pounds of this copper scale and only in a few hours. But in the annealing chamber, near the 'eye' of the furnace, you should leave it alone for 4 days, then tumble and grind it very well, passing it through a fine sieve.

Now return it to the terracotta oven pans, or tiles, as before with the same fire and heat for 4 days. It will turn to a black powder and will lump together. Grind and sift it through a fine sieve and again put it back on tiles and heat for 4 days. At this point, the copper scale will be optimally prepared without the annoyance and expense of the former method, and all in all, it will have the same effect in coloring. Take care to wash it thoroughly of all sediment before calcining it. As they say: The sign will be that it makes the glass swell and boil, only then is it well prepared.

~ 29 ~
A Fine Crystal in Aquamarine with the Above Copper Powder

To a large crucible with 60 pounds of crystal frit that has been thoroughly de-salted, as described before, and not thrown into water, because as I stated when I made the aquamarine it is my custom never to throw the crystal in water. It seemed to me that it came out better that way. However, you can try it both ways and adopt the one that works best for you.

In any case, into the 60 pounds of crystal, when it is very well clarified, put 1½ pounds of the copper scale, made at reduced expense as described above, and with it mix in 4 ounces of prepared zaffer. Mix these two powders well, then add them to the crystal in four parts. Stir the powder into the glass well for 2 hours, then turn and blend it well, as per usual. Do a test and if the color appeals to you leave it since there is no need to re-tint for 24 hours. Then turn and blend the glass again, so that the color becomes fully uniform. Working it will make the finished pieces come out better, and this will make a most beautiful aquamarine, as I have made many times with blessed success. You must charge it with more or less color in accordance with its intended use, but above all take note that for beadmaking cane you must apportion the calcined copper scale by mixing it into frit that is half rocchetta and half crystal. You will have beautiful aquamarine in either case.

~ 30 ~
A Lower Cost Aquamarine

Use the same copper scale, prepared as before. To Levantine rocchetta, or aternately to Spanish [barilla] add zaffer in the prescribed doses, in the same way and form discussed previously. Make sure that none [of the melts] have had manganese added and that they are very well de-salted, but not thrown in water. Use the rules that were discussed before for the crystal. In this way, you can make a very beautiful aquamarine for quick jobs that will cost much less than crystal, because the rocchetta salt costs much less than the bollito, as is well known. I have made it this way in Pisa many times and always with good success.

26   LIBRO PRIMO
traggiettato in acqua: perche, come ho dichiarato quando faceuo l'acqua marina, non coſtumauo mai tragiettare in acqua il criſtallo, & mi pareua, che veniſſe meglio: però in queſto ſi può prouare nel vno, & nel altro modo, & attenerſi al meglio, adunque a dette libre ſeſſanta di criſtallo, quando è pulito beniſſimo, ſi dia libbre vna, & mezzo di detta ramina, fatta con manco ſpeſa, come ſopra, & con eſſa ſi meſcoli oncie quattro di zaffera preparata, meſcolando bene inſieme queſte due poluere, poi ſi dieno al detto criſtallo in quattro uolte, meſcolando bene la poluere con il vetro per duoi hore, poi ſi torni a rimeſtare bene al ſolito, & ſe ne caui vna proua, & ſe il colore piace ſi laſci coſì non hauendo biſogno di caricare per ven tiquattro hore, poi ſi torni a meſcolare il vetro di nuouo, acciò il colore venga bene vnito, & ſi lauori in quei lauori, che più guſtano, & queſta farà vn'acqua marina belliſsima, come ho fatto io molte uolte con felice ſucceſſo. Il caricar la più, e manco ſi deue fare ſecondo i lauori, a che deue ſeruire, però ſopratutto ſi deue auuertire, che la ramina ſia calcinata a proportione per canne da conteria ſi meſco lerà la metà di fritta di rocchetta con fritta di criſtallo, che farà acqua marina bella in ogni modo.

Acqua marina di manco ſpeſa. Cap. XXX.

L A medeſima ramina preparata, come ſopra con la detta doſi di zaffera, ſi dia nel medeſimo modo, & forma alla rocchetta di Leuante, & anco a quella di Spagna, che neſſuna di loro habbi hauuto manganeſe, & che ſiano beniſ ſimo deſalate, peró non tragietare in acqua, vſando le regole, che ſopra ſi dicono nel criſtallo, che in queſta maniera ſi farà vn'acqua marina aſſai bella per ogni forte di lauori, & coſterà manco aſſai del criſtallo, perche la rochetta vale manco aſſai del bollito, come è noto. In queſta maniera l'ho fatta io molte volte in Piſa, & ſempre con ſuc ceſſo buono.

Acqua

~ 31 ~
A Marvelous Aquamarine Beyond All Aquamarines, of My Invention

The dry distillation of vitriol of Venus [copper sulfate] chemically made without corrosives, if left to stand in the open air for some days will take on, without any intervention, a pale green color. Pulverize this material accompanied by prepared zaffer in the same doses as were discussed in the other prepared copper scale recipes. Add it to the crystal in the way and manner laid out for the other aquamarines. This will make a beautiful aquamarine so nice and marvelous, that you will be astonished, as I have done many times in Flanders in the city of Antwerp to the marvel of all those that saw it.

The way to make the dry distillation of copper vitriol without corrosives, spagyricaly [naturally], is to take small pieces of thin copper, the size of a half a florin and have one or more large crucibles according to your need. In the bottom of them, put pulverized common sulfur from the earth [brimstone] and over that put the pieces of copper discussed above. Then add more pulverized brimstone and over it pieces of copper. Continue in this manner until all the copper is put to work. To undertake this procedure is otherwise known as 'stratification'. Follow the instructions for this laid out in chapter 131, and then test it. To your great contentment, you will be astonished at what you see. I do not know of anybody else who has tried it this way and I Priest Antonio Neri trying it found it most marvelous, as said above, and it is of my own invention.

DELL'ARTE VETRARIA   27

Acqua marina marauigliofa fopra tutte l'acque marine di mia inuentione.   Cap. XXXI.

IL capo morto del fpirito di vitriolo di Venere chimicamente fatto fenza corrofiui lafciato ftare all'aria per alquanti giorni, piglia per fe medefimo fenza niuno artifitio vno colore verde sbiadato. Di quefta materia poluerizzata con l'accompagnatura della zaffera preparata, & con la medefima dofi, che nelle altre ramine preparate fi è detto, dato al criftallo nel modo, & forma detta nelle altre acque marine, farà vna acqua marina tanto bella, & marauigliofa, che farà cofa di ftupore, come io ho fatto più volte in Fiandra nella città di Anuerfa con marauiglia di tutti quelli, che l'hanno vifta. Il modo di fare il Vitriolo di rame fenza corrofiui, fpagiricamente farà pigliare pezzetti di rame fottili, della grandezza di mezza piaftra Fiorentina, & hauere vno, ò più correggiuoli fecondo il bifogno, & nel fondo di effi mettere vno fuolo di zolfo comune poluerizzato, & per fopra delli pezzetti di rame fopradetti, & poi vn fuolo di zolfo poluerizzato, & per fopra pezzetti di rame, & in quefta guifa operare fino tutto il rame fia meffo in opera, che si hauerá prefo per quefto effetto, che altrimenti quefto fi dice ftratificare: fatto quefto, fi quoprino, fegue in quefto al Cap. 140. acciò lo poffino prouare, che con loro contento vedranno cofe di ftupore, quefto modo non sò che neffuno l'habbi prouato, & io Prete Antonio Neri prouandolo, lo trouai marauigliofo, come fopra; però lo dico di mia inuentione.

Verde fmeraldino in vetro.   Cap. XXXII.

NEL fare il verde, fi habbi confideratione, che il uetro non habbia molto fale perche pigliando uetro, che habbi affai fale, come è bollito; & la rochetta, non ui fi può fare uerde bello, ma fi bene l'acqua marina; perche il fale confuma il uerde, & fempre tira il colore all'azzurigno, & marino. però quando fi uuole fare un bel uerde habbifi in padella, ò padellotto uetro comune, cioè fatto di poluerino, come fi è moftrato al Capitolo ottauo, & que-

D   2   fto

~ 32 ~
Emerald Green in Glass

In making green take into consideration that the glass should not have a lot of salt in it.

If you use glass that does have a lot of salt, like bollito and rocchetta, you cannot make a beautiful green but only aquamarine, since the salt consumes the green and it always inclines to a bluish marine color.

Therefore, when you want to make a beautiful green, in a large or small crucible put common glass made with polverino, as can be seen in chapter 8. Use glass that has never, in any manner, been shown manganese, because then your work would appear dark and ugly. Once it is fused and very well clarified, add a little crocus of iron to this glass, which has been calcined with vinegar, as is described in chapter 17. So as an example, for 100 pounds of glass add approximately 3 ounces [of crocus]. Leave the glass for an hour, then stir well until it incorporates the crocus pigment, which makes the glass become somewhat yellowish and removes the scum and bluishness that glass always has. This work will give the glass a beautiful green tint at which point add the 'thrice baked' copper made of scale from the hammerings of kettle-smiths, as shown in chapter 28. For this the rule is to use 2 pounds for every 100 pounds of glass, added in six portions. Mix the powders well with the glass, and then leave it for 2 hours to settle down and incorporate into the glass.

At this point return to mix the glass and see if your color is sufficient, and in compliance for its intended use. You may need to add more copper scale to produce the color, and you may need to charge it more or less according to the work and use. If you need to make some adjustments, I cannot give you precise rules. If the green inclines to the marine and bluish and it does not appeal to you, add a little bit more crocus of iron, as above, to obtain a most beautiful emerald green also known as leek green. At the end of 24 hours it will be ready to work. Be ever vigilant; before the glass is worked make sure to stir it, because the colors always go to the bottom and on top the crucible is often more [color] free. I made this green in Pisa and it always came out quite beautifully and with such success every time, as it will for you; just follow the instructions step by step, as I keep saying.

~ 33 ~
A Nicer Green than Above

Now, if you want to have a much nicer luminous green than the one above, take a crucible of crystallino that has not had manganese and has been thrown in water once or twice so that all saltiness is purged. To this crystallino, add half again that amount of common white glass made with polverino, which has not had any manganese. This mixed glass should be fused and clarified well. For every 100 pounds of it, add 2½ pounds of 'thrice baked' scale made with copper shims in the arches of the furnace, as was shown in chapter 25. With this, unite 2 ounces of crocus of iron calcined with sulfur and broken down, as demonstrated in chapter 17 [sic. 16]. Unite these two powders thoroughly; add them to the glass using the rules for the above green.

You may need to charge the color more or less according to your judgment and if the glass has a bluish tint, add a little of the above crocus, which will go to work on it and make it diminish, like the other green. Then it will be a wonderful pimpernel green. I have made it this way many times in Pisa with good success for jobs more exquisite than ordinary. However, you must always remember; the copper must be well prepared if you want

28   LIBRO PRIMO

fto uetro non habbia manganefe di niuna maniera, perche l'opera verrebbe nera, & brutta, come è fufo, & pulito benifsimo; fi dia a quefto vetro vn poco di Croco di ferro, cioè di quello, che è calcinato con l'aceto, come fi moftra al capitolo decimofettimo, & di quefto per efempio fe ne dia à libbre cento di vetro oncie tre in circa, fi mefcoli bene il vetro, & fi lafci per vna hora, tanto, che il vetro incorpori la tintura di detto croco, che farà venire il vetro alquanto giallogniolo, & gli lieua il rozzume, & azzurringo, che fempre ha il vetro, che opera poterfi dare al vetro vn bel verde all'hora s'habbi la ramina di tre cotte fatta di fcaglia, & battitura de i calderai, come fi è moftrato al Capitolo ventotteefimo, & di quefta fe ne dia à ragione di libbre dua per ogni cento libbre di vetro, & quefta fi dia in fei uolte mefcolando bene le polueri con il vetro, poi fi lafci pofare, per duoi hore, & incorporare il vetro, & all'hora fi ritorni a mescolare il vetro, & fi vegga fe è carico a baftanza, conforme al lauoro, che ha da feruire, & bifognando, fe li può dare più ramina, & in effetto i colori fi hanno a fare più carichi, & meno carichi fecondo l'opera, & lauori, che fe ne deue fare, che in quefto non fe ne può dare regola certa, & fe il verde tendeffe al marino, & azzurigno, & non piaceffe, fe li può dare vn'altro poco di croco di ferro, come fopra fi è detto, & cofi fi hauera verde bellifsimo fmeraldino, altrimenti detto verdeporro, che in capo di ventiquattro hore fi potrà lauorare, auuertendo fempre mai, quando fi deue lauorare, per prima mefcolare il vetro, perche i colori danno fempre in fondo, & in cima della padella fono fempre più fcarichi. Quefto verde moltifsime volte lo feci in Pifa, & fempre mi venne affai bello, & tale riufcirà a ciafcuno, che offeruerà puntualmente quanto fopra fi è detto.

Verde più bello del fopradetto.   *Cap. XXXIII.*

MA fe volefsi hauere vn verde molto più bello del fopradetto, ò più lampante, habbi in padella criftallino, che non habbi hauuto manganefe, & fia tragiettato in acqua per vna, ò dúa volte, tanto che ogni falfedine gli efca di doffo, & à quefto criftallino in partita li fia dato

to have beautiful colors.

~ 34 ~
A Marvelous Green

You should have the 'thrice baked' copper scale, made of hammerings and flakes of kettle-smiths, as has been shown [chap. 28]. Instead of using crocus of iron made the many ways shown above, take the very fine iron scales that fall from the anvils of ironsmiths. Clean them of charcoal, dust, and cinders. Grind and sift finely, and with the doses described above, mix them very well with the calcined copper scale and add them to common polverino glass that has not had manganese, as per the above rules for green. With this crocus of iron, or true scale, you will have, without a doubt, a marvelous emerald green. It will have completely lost the bluish sea color, that ordinarily every glass has. Instead it will make a yellowish green to emerald green and have the most beautiful shine and luster; more than any of the above greens.

To put the iron scale in with the copper scale was invented by me, Priest Antonio Neri. In the rest, observe the doses and rules as described above for the other greens. If you do this, you will have a thing of astonishment, as frequent experience has shown me.

DELL'ARTE VETRARIA. 29

to la metà di vetro comune bianco, fatto di polueri-no, che non habbi hauuto manganese; come questo ue-tro mescolato è fuso, & pulito bene, piglia per ogni libre cento di esso libre dua, e mezzo di ramina di tre cotte fatta con lamine di rame nelli archi della for-nace, come si è mostrato al Capitolo vigesimoquinto, & con questa vnisci oncie dua di croco di ferro calcinato con zolfo, & reuerberato, come si è detto al suo luogo nel capitolo decimosesto : vnite bene insieme queste due poluere, le darai al sopradetto vetro, vsando le regole di sopra nel verde dette. Potrai caricare il colore più & man co secondo ti pare, & se il vetro haueste del azzurigno, dagli vn poco di Croco sopradetto, che glielo fa perdere, & lauorarlo, come l'altro verde, che farà verde di pimpi-nella stupendo. Io l'ho fatto in questa maniera molte volte in Pisa con buono successo per lauori più esquisiti del li ordinari: però si auuerta sempre mai, che le ramine sie-no ben preparate, se si vuole hauere colori belli.

Verde marauigliofo. Cap. X XIII.

HABBISI la ramina di tre cotte, fatta di battitura, & scaglie di calderai, come si è mostrato: poi in cambio di Croco di ferro fatto come sopra in tanti modi, si pigli scaglia di ferro, che cade dalla incudine de i fabbri ferra-ri, & questa pulita da carbone, poluere, & cinigie si pesti, & stacci sottilissimamente, & con la dosi sopradetta, si me-scoli con la ramina benissimo, & si dia al vetro comune di poluerino, che non habbi hauuto manganese, con le re-gole sopradette nel verde, che con questo croco di ferro, ò vero scaglia senza dubbio si hauerà vn verde smeraldi-no a marauiglia, quale in tutto, & per tutto hauerà perso il colore azzurigno, & di mare, che per ordinario ha ogni vetro, ma farà vn verde gialleggiante alla smeraldina, & & hauerà vn lampante, & lustro bellissimo più delli so-pradetti verdi: il mette gli la scaglia di ferro con la rami-na fu inuentione di me Prete Antonio Neri; Nel resto si osseruino le dosi, & regole sopradette nelli altri verdi, che si hauerà cosa di stupore, come l'esperienza più volte mi ha mostrato.

Altro

~ 35 ~
Another Green, Which 'Carries the Palm' for All Other Greens, Invented by Me

To a large crucible, add 10 pounds of glass, which is half crystallino, thrown in water several times, and half common white polverino glass [and fuse them]. Now take 4 pounds of common polverino frit, with this, mix in 3 pounds of red lead [lead oxide]. Unify them well and add them into the above crucible. In a few hours everything will have clarified, now purify it by throwing it in water. Inspect the glass carefully before returning it to the crucible.

All lead precipitating out of the glass must be removed with diligence, throwing it away, so that it does not make the bottom of the crucible break out, as can happen. Return the glass that was thrown in water to the crucible and leave it to clarify for a day. Then add the color using the powder made by dry distillation of vitriol of copper [chapter 31] made chemically. Also, add a little crocus of iron, but very little. The result will be a most marvelous beautiful green, the best that I ever made. It will seem just like an emerald of ancient oriental rock, and you can use it in every sort of job.

~ 36 ~
Sky Blue, or More Properly Turquoise, a Principal Color in the Art of Glassmaking

Take the sea salt known as black salt, or rather coarse salt, since the ordinary white salt, that they make in Volterra would not be good. Put this salt in a frit kiln or oven to calcine, in order to release all moisture, and turn white. Next, grind it well into a fine white powder. This salt now calcined should be stored for the use of making sky blue or rather turquoise color as described below.

DELL'ARTE VETRARIA.   31

rino calcinato mefcolandolo bene con il uetro, come fi fa la ramina, & di quefto fe li dia a poco a poco fino, che il colore d'acqua marina perda il trafparente, & diafano, & pigli l'opaco, perche il fale vetrificandofi fa perdere al vetro il trafparente, & gli dà vn poco di sbiadato, & cofi a poco a poco fa il colore detto aierino nell'arte vetraria, che è il colore della pietra detta turchina, colore principalifsimó nell'arte, quando il colore fta bene, bifogna lauorarlo fubito, perche il fale fi perde, & fuapora, & torna di nuouo il vetro trafparente, & con brutto colore: anzi quando nel lauorare perdeffi il colore fe li da nuouo fale bruciato, come fopra, acciò il colore ritorni, & cofi fi ha il defiderato colore. Auuerta qui bene il conciatore di fornace, quando dà quefto fale, che quando non è bene calcinato, fempre fcoppia, però fia cauto, & gnardi gl'occhi, & il vifo, perche v'e pericolo di farfi del male, la dofi del fale farà darlo a poco a poco mettendo alquanto di tempo in mezzo da vna volta all'altra, tanto che fi vegga il defiderato colore, che in quefto io non vfauo altra dofi, e pefo, folo con l'occhio quando vedeuo, che il vetro era al defiderato fegno di colore, non gli dauo più fale, che tutto fta nella fperienza, io faceuo quefto colore fpeffo, perche è molto neceffario nella conteria, & il più ftimato, e pregiato colore, che fia nell'arte. Però per fare aierino per conteria fi pigli acqua marina fatta di metà di bollito, & metà di rochetta, che verra colore bello, etiam che non fia tutto bollito.

You should have bollito frit also known as crystal in a large or small crucible. It should be tinted aquamarine color, for which I have previously given many recipes. Let the color be beautiful and well charged. This is of great importance in order to make a beautiful sky blue. So accordingly, you should start with a beautiful and optimal aquamarine. To this colored glass then, add the above calcined sea salt stirring it well into the glass, as was done with the copper scale, add it little by little, until the aquamarine color loses its transparency and diaphany becoming opaque.

The vitrified salt makes the glass lose its transparency and gives it a slight paleness, so that little by little you will make the color that is called sky blue in the art of glassmaking. This is the color of the stone called turquoise, a color foremost in the art. When the color is good, you must work it quickly, because the salt will be lost and evaporated and the glass will become transparent again, with an ugly color. Indeed, if you start to loose the color while working, add new roasted salt, as before, so that the color returns and therefore you will retain the desired color.

Furnace compositioners should take careful note here, when you add this salt, if it is not well calcined it always bursts. Therefore, you should be cautious and shield your eyes and vision, because there is a danger you could be hurt. Add the doses of salt little by little putting in a bit at a time pausing from one time to the next until you see the desired color. With this, I do not rely on either dose or weight, but only on my eyes. When I see that the glass reaches the desired level of color, I stop adding salt. This all comes with experience; I have made this color often, because it is very necessary in beadmaking and is the most esteemed and prized color in the art. In order to make sky blue for beadmaking take aquamarine made with half bollito and half rocchetta glass, and it shall be a beautiful color, even though it is not all bollito.

GLOSSARY

Appearing below are specialized terms Neri uses throughout the book. These terms are listed both in the original Renaissance Italian, and in their equivalent modern English translations. In each case, the English entries contain references leading you to the original Italian terms, where you will find the appropriate descriptions.

Acqua forte - literally strong water. Nitric acid (HNO_3)

Acqua regis - literally king water. A mixture of Nitric and hydrochloric acids Various proportions were used, depending on the material to be dissolved. Commonly, more nitric acid than hydrochloric was employed.

Alkali salt - see Sale alchali

Ammoniac salt - see Sale ammoniaco

Ammonium chloride - see Sale ammoniaco

Annealing chamber - see Era

Aqua fortis - see Acqua forte

Aqua regia - see Acqua regis

Artificial crystal - see Bollito

Assay - see Gustare

Balas ruby - see Balascio

Balascio - balas ruby, a ruby of a delicate rose-red variety. From Persian Badaksan, a district of Afghanistan where it is found.

Barilla - synonym for 'soda', the Spanish term for plant ash rich in alkali salts.

Beadmaking - see Conteria

Blue iris - see Fioralisi

Blue magpie - see Gazzera marina

Bollito - glass frit used specifically for higher quality crystal glasses.

Borage - a herbaceous plant with bright blue flowers and hairy leaves. Borago officinalis from the Latin 'vorago', and perhaps from Arabic 'abu huras' = 'father of roughness' (referring to the leaves).

Brazilwood - see Verzino

Brimstone - see Suolo di zolfo

Broom flower - see Ginestra

Cabbage flower - see Fior cappuccio

Calamine - see Zelamina

Calcara - an oven used to 'calcine' or roast materials used in glassmaking.

Calcinare - to calcine, reduce, oxidize, or desiccate by roasting or strong heat. The term comes from the roasting of calcium carbonate ($CaCO_3$) in order to form lime (CaO), a process invented before the first century B.C. by the Romans to make cement.

Calcine - see Calcinare

Calcium oxide - see Sale della calcina

Chrysolite - see Grisopatio

Common glass - see Vetro commune

Compositioner - see Conciatore

Conciatore - glass compositioner. One who makes glass from its constituent materials, adds colorants, and supervises the maintenance and use of the glass melt.

Conteria - the craft of beadmaking. Although there are many references to beadmaking throughout the book, this term was apparently unknown to Neri's first translator, Christopher Merrett, who transcribed it as 'counting house'. Therefore the term is absent or garbled in subsequent translations, which were all based on Merrett.

Copper sulfate - see Vitriol of venus

Coreggiolo - also correggiuolo, crogeolo, crociolo, a melting pot or small hand-sized crucible sometimes used by goldsmiths, as opposed to the much larger padella and padellotto, which were used for the glass melt.

Correggiuolo - see Correggiolo

Cristallino - a medium quality glass midway between common glass and crystal.

Cristallo - the best quality glass, exceptionally clear and workable.

Cristallo di montagna - mountain or rock crystal; a naturally occurring form of exceptionally clear quartz. This term is also used by Neri to refer to especially fine crystal glass; '*artificial rock crystal*'.

Croco de ferro - synonym, see Croco de marte

Croco de marte - literally saffron of Mars, finely ground rust made from oxidized iron filings.

Crocus martis - see Croco de marte

Crocus of iron - see Croco de marte

Crucible - see Coreggiolo, Padellotto, Padella.

Crystal - see Cristallo

Crystallino - see Cristallino

Cupric sulfate - synonym for copper sulfate, see Vitriolo di venere

Decant - see Decanti

Decanti - decant, to gradually pour from one container into another, typically in order to separate the liquid from the sediment. From Latin *de-* 'away from' + *canthus* 'edge, rim'.

Desalato - de-salt, the process by which impurities, especially glass gall, is skimmed off a crucible of molten glass in the furnace.

De-salt - see Desalato

Destillatoria - distillation; the art of extracting essential oils etc. from plants and herbs, used in various medicines and remedies.

Era - annealing area or chamber of a glass furnace, a spot where finished work was allowed to cool slowly to avoid stress cracking. Depending on the design of the furnace, this was sometimes in the same chamber as the main glass crucibles, sometimes in another chamber, and sometimes in a separate structure all together. In english was called the lehr, lear, or leer.

Esau - a biblical character, pronounced 'sw+hairy' was the oldest son of Isaac and Rebecca, and twin brother of Jacob.

Eye - see Occhio

Faua - a broad bean rich in potassium, the fava bean is native to the Mediterranean basin. It is a favorite around the world. The pods are 6 to 12 inches long and the tip of one end is pointed. Inside, there are 5 to 10 flat, round ended seeds which vary in color from green to red to brown to purple. A member of the *Vicia* genus, also known as Habas, Broad bean, Faba bean, Horse bean, English bean, Windsor bean, Tick bean, Cold bean, and Silkworm bean.

Fava bean - see Faua

Ferretto - iron filings.

Fiandra - Flanders, a country that covered the northern half of Belgium, and included parts of France and southern Holland. Neri spent the years between 1604 and 1611, in the Flemish city of Antwerp, sponsored by his friend Emanuel Ximenes, and working in the glass shop of Filippo Ghiridolfi. During this period the country was in the process of regaining independence from Spain. The name Flanders probably is of Celtic origin, derived from the term for 'swampy region'.

Fior cappuccio - a day lily that bears large yellow flowers, each lasting only one day. Genus *hemerocallis*.

Fioralisi - blue iris; '*fiore+aliso*' = blue lily-flower (a variety of iris).

Flanders - see Fiandra

Florin - from Italian *fiorino*, diminutive of fiore 'flower' originally a Florentine coin bearing a fleur-de-lis minted between 1252 and 1553. It was a little larger than a US Quarter (a little over an inch in diameter), and contained 3.54 grams of gold. Adopted in various forms throughout Europe, notably in England, France, Austria, Poland, Germany, Flanders, and Dalmatia.

Frit - see Fritta

Frit kiln - see Calcara

Frit rake - see Riauolo

Fritta - frit, a mixture of silica and fluxes which is fused at high temperature to make glass. See Neri's description of calcination in chapter 8.

Furnace Compositioner - see Conciatore

Gazzera marina - the Iberian bird 'gazza azzurra' or blue magpie; cyanopica cyanus. It has an intense deep blue plumage. Not to be confused with the *gazza marina*, or razorbill, a relative of the penguin also frequents the Italian coast, but sports only black and white plumage.

Ginestra - broom, a shrub typically having many yellow flowers, long, thin green stems, and small or few leaves. *Cytisus scoparius* and many other species, especially in the genera *Cytisus* and *Genista*. The stems were commonly bundled together to fashion various brushes for sweeping.

Girasol - a kind of opal reflecting a reddish glow when held up to the light. Derived from '*Gira*': to turn + '*sol*': the sun.

Giunchi - rush or bull rush, a marsh or waterside plant with slender stem-like pith-filled leaves, some types are used for matting, baskets, etc. Genus *Juncus*.

Glass - see Vetro

Glass Compositioner - see Conciatore

Glass gall - see Sale alchali

Golden day lily - see Fior cappuccio

Goslin weed - see Robbia

Grisopatio - the mineral olivine, also known as chrysolite. The choicest green stones used as gems are called Peridot. The chemical composition is usually given as $(Mg\text{-}Fe)_2\text{-}SiO_4$, but is really a solid solution of forsterite, $2MgO\text{-}SiO_2$ and fayalite, $2FeO\text{-}SiO_2$

Gruma - see Tartaro

Gustare - literally to taste, the process of assaying, or testing a small quantity of material to ascertain its quality or suitability for a given purpose.

Herbal Distillation - See Destillatoria

Homeopathy - see Spagirica

Kali - *salsola kali*, also known as saltwort, tumbleweed, and Russian thistle. A shrub bush high in alkali content, common throughout Europe and the United States. Uses in the ancient world included heating, soap and glass production. The ash contains 20% K, 18% Ca, 3% Mg, 1.5% Al, 1.5% Fe, 6% phosphate, 6% sulfate, 40% carbonate, and 2% chloride (*List and Horhammer, 1969-1979*).

Kiln - see Calcara

Leek green - see Verdeporro

Leuante - the Eastern Mediterranean region around modern Syria and Lebanon, extending to Greece, Turkey, Israel, and Egypt. 'Levant' loosely translates from French to 'land of the rising sun'.

Levant - see Leuante

Libra - also *libbra*; pound, an old Roman unit of weight used in Renaissance Florence. In current units it is (339 grams), almost equivalent to ¾ of a modern avoirdupois pound. Not to be confused with the Troy pound, which is 373.2 grams. A Libra consisted of 12 'onces' - almost the same ounces we use today.

Lime - see Sale della calcina

Lime salt - see Sale della calcina

Lutare - lute: To coat with liquid clay or cement to seal a joint, coat a crucible, or protect a vessel from the direct heat of the furnace. From the latin. *lutum* 'potter's clay'.

Lute - see Lutare

Lye - see Ranno

Manganese - an elemental metal (Mn) that in powder form is a silvery-white with purplish shades. Used in glass as a de-colorizing agent.

Millet - see Saggina

Minio - 'minimum' also known as red lead. Lead tetroxide (Pb_3O_4), a bright red or orange-red powder, insoluble in water. As a pigment it has great covering ability and brilliance. Added to the glass melt it becomes colorless. Not to be confused with another brilliant red pigment called vermillion or cinnabar, which is mercury sulfide, HgS. In earlier times it too was referred to as 'minimum'.

Nitric acid - see Acqua forte

Occhio - the 'eye' of the furnace or annealing chamber. This was a hole connecting the combustion chamber where wood was burned to the glassworking areas of the furnace, essentially a heat vent.

Oltramarino - an ancient brilliant deep blue pigment made by grinding lapis lazulai. From (*azzurro*) oltramarino, lit. 'azure from overseas' (because the lapis lazuli was imported).

Once - ounce, a unit of weight almost equivalent to a modern avoirdupois ounce. In current units it is 28.25 grams. In Renaissance Florence, the Roman system of weights was used in which a *libra* (pound) was divided into 12 *onces*, not 16. Not to be confused with Troy weight. (see Libra)

Opposition - an astronomical condition where the earth (in this case) is directly between the moon and the sun; in other words, when the moon is high at midnight.

Orbio - see Occhio

Orpello - gilding or gold paint, however Neri uses the term to describe thin foil strips or tinsel made from copper sheet treated with calamine to form a gold colored brass composition. In chapter 20 he advises saving money by using old wreaths and garlands made from it for a nice blue.

Orpiment - see Orpello

Ounce - see Once

Padella - the smaller of the large crucible pots used inside the furnace to hold molten glass.

Padellotto - a larger version of the padella, a large crucible used to hold up to several hundred pounds of molten glass in the furnace.

Palm - see Porta la palma

Peridot - see Grisopatio

Pimpernel - see Pimpinella

Pimpinella - pimpernel, a low-growing plant with bright five-petalled flowers. Anagallis arvensis (scarlet pimpernel) is one species. From Latin piper 'pepper'. In chapter 33, Neri's reference to 'a wonderful pimpernel green' is presumably alluding to the plant's leaves.

Plinio - (the elder) Gaius Plinius Secundus, a prolific Roman first century author and historian, most noted for his 37 volume *Natural History*.

Pliny - see Plinio

Poluerino - plant ash for glassmaking sold in the form of a powder, hence the name derived from 'Polvere' or powder. See also Rocchetta.

Polverino - see Poluerino

Porfido - porphyry a hard igneous rock containing crystals of feldspar in a fine-grained, typically reddish groundmass. Used for grinding and sharpening. From the Greek porphura 'purple'.

Porphyry - see Porfido

Porta la palma - literally 'carries the palm'. An expression used by Neri in the title of chapter 35 to describe his very best green. A reference to the custom of Catholic clergymen carrying palm fronds and leading the Palm Sunday procession, in observance of Jesus' entry to Jerusalem, where his followers laid palm leaves at his feet to walk on. However, the custom of victorious armies carrying palms in a procession through the vanquished territory is one that dates back at least to the early Roman Empire, and probably much earlier.

Pound - see Libra

Prince of Orange - see Principe d'Arangie

Principe d'Arangie - Maurice, Prince of Orange (1567 - 1625), A brilliant military strategist was instrumental in regaining Flanders' independence from Spain. Technically, his title was Prince of Nassau, until the death of his brother Philip William in 1618. A direct ancestor of the current Dutch Queen Beatrix. He surrounded himself with artists, craftsmen and intellectuals.

Rake - see Riauolo

Rame - metallic copper

Ramina - copper scale, flakes of copper made as a byproduct of the smithing process.

Ranno - a strong alkali solution called lye, in this case made by boiling vegetable ash.

Red lead - see Minio

Regis - see Acqua regis

Riauolo - a frit rake, as Neri describes it in chapter 2 "the riauolo is an instrument of iron, very long, with which one agitates the frit continuously". His instrument is still very well known in the glass making houses. For the Muranese it remained 'reàulo' until the end of the 1800s: Currently it is known as a 'reauro'.

Robbia - in Neri's time the second most important source of dye, after indigo. Rubia tinctorum also known as madder root, Turkey red, and alizarin. Its alkaline solution was used with various mordants to give madder red (with aluminum and tin), blue (with calcium), and violet-black (iron). Purpurin purple dye was also produced with it.

Rocchetta - plant ash for glassmaking sold in the form of large pieces, hence the name derived from 'rocca' or rock. See also Polverino.

Rock crystal - see Cristallo di montagna

Roman pound - see Libra

Rush - see Giunchi

Saggina - a grass, or cereal which bears a large crop of small seeds, used to make flour or alcoholic drinks. Includes Panicum miliaceum and other species.

Saint Jerome - see San Girolamo

Sale - salt, in this context Neri is referring to the vegetable salts extracted from plant ash, mostly potassium and calcium compounds. In general it is any chemical compound formed by the reaction of an acid with a base, with the hydrogen of the acid replaced by a metal or other cation.

Sale alchali - alkali salt, also known as 'glass gall', sandever, and sandiver. The whitish salt skimmed from the surface of melted glass, which is cast up, as a scum, from the fusion process.

Sale ammoniaco - ammonium chloride (NH_4Cl). Added to nitric acid to form acqua regia.

Sale della calcina - lime salt, calcium oxide (CaO), also known as 'quicklime' a caustic alkaline white powder made by heating limestone ($CaCO_3$). Used in the

manufacture of cement for the construction trades.

Salt - see Sale

Saltwort - see Kali

San Girolamo - controversial and prolific late fourth century theologian. Wrote a revision of the Latin version of the Book of Job in the year 384. Author of the Vulgate translation of the Old Testament. Reputed to have removed a thorn from a lion's paw, thereby winning its devotion for years. Also known as Saint Jerome, Eusebius Hieronymus Sophronius, and Hieronymus.

Saxon blue - see Zaffera

Soda - used by Neri as a generic term for plant ash from a variety of sources. From the Arabic term suwwad meaning 'saltwort'(kali). In current usage it refers to Sodium carbonate.

Soda di Spagna - plant ash for glassmaking from Spain.

Spagirica - The art of reducing things to their constituent elements (or essences), purifying them, then reconstituting them. This was done by a variety of means including solution, distillation, and evaporation.

Spagyric - see Spagirica

Spanish barilla - see Soda de Spagna, and Barilla.

Spanish soda - see Soda di Spagna

Sulfuric acid - see Vitriol

Suolo di zolfo - Sulfur from the ground, or brimstone.

Tarso - quartz pebbles of exceptionally pure silica (SiO_2) found in riverbeds. The Muranese preferred that of the Ticino River in Pavia.

Tartar - see Tartaro

Tartaro - also known as gruma and argol. The reddish incrustation which forms on the inside walls of aged red wine barrels consisting of yeast mixed with potassium bitartrate. Chemically it is a potassium compound formed through a reaction with tartaric acid, a major constituent of grape juice. Pure tartar takes the form $KHC_4H_4O_6$.

Test - see Gustare

Tiberio - Tiberius Claudius Nero 42BC to 37AD followed Augustus as emperor of Rome. Pliny and Petronius recorded the story of an artisan who upon discovering a flexible glass was put to death by Tiberius in order to protect the value of his gold and silver holdings.

Tiberius - see Tiberio

Tinsel - see Orpello

Trag[i]ettare - a term Neri uses for the process of throwing or flinging molten glass into vats of water as a means of removing excess salt, glass gall, and other impurities.

Ultramarine - see Oltramarino

Verdeporro - a bright green color of the of the leek, a plant related to the onion, with flat overlapping leaves forming an elongated cylindrical bulb that together with the leaf bases is eaten as a vegetable. Allium porrum.

Vermillion - see Minio

Verzino - the wood of the Brazilian trees Caesalpinia brasiliensis, C. crista, and C. echinata, used for dyes. It produces purple shades with chrome mordant, and crimson with alum. The wood is prized for fine furniture and violins, it has a rich bright-red color, and takes a fine lustrous polish.

Vetrificare - to convert into glass or a glass-like material through the application of heat.

Vetro commune - common glass, the most basic and lowest quality glass.

Vitrification - see Vetrificare

Vitriol - See Vitriolo

Vitriol of copper - see Vitriolo di venere

Vitriol of venus - see Vitriolo di venere

Vitriolo - sulfuric acid (H_2SO_4).

Vitriolo di rame - synonymous with 'Vitriolo di venere', *venere* or 'venus' was an alchemist's term for copper.

Vitriolo di venere - cupric sulfate ($CuSO_4$) also called bluestone. Naturally occurring as Chalcanthite.

Zaffer - see Zaffera

Zaffera - Saxon blue, a mineral mixture produced in Saxony. A deep-blue powder made by fusing cobalt oxide with silica and potassium carbonate, It contains 65 to 71% silica, 16 to 21 potash, 6 to 7 cobalt oxide, and a little alumina.

Zanech - Neri sites him as the father of Job according to Saint Jerome. This is not supported in current biblical scholarship. Possibly a reference to Hanoch, grandson of Abraham & Keturah (Abraham-Midian-Hanoch) as distinct from the Abraham & Sarah blood line (Abraham-Isaac-Esau).

Zelamina - calamine a mineral ore occurring in crystal groups with a vitreous luster. It may be white, yellowish, greenish, or brown. Chemically it is a zinc silicate ($2ZnO \cdot SiO_2 \cdot H_2O$) usually only containing about 3% zinc. It was concentrated through roasting and distillation, and then mixed directly with copper to form Brass. (Not to be confused with the soothing pink ointment called calamine, which is made from a mixture of zinc carbonate and ferric oxide.)

"... through experience one can discover and learn much more than could ever be accomplished through long study."

- Antonio Neri

SELECTED BIBLIOGRAPHY

Abbri, Ferdinando, *Antonio Neri, L'Arte Vetraria.* Florence: Giunti, 2001, ISBN 88-09-01267-4 (paper). Italian reprint by Neri's original publisher, with updated spelling and typography.

Ball, Philip, *Bright Earth, Art and the Invention of Color.* New York: Farrar, Straus, and Grioux, 2001, ISBN 0-374-11679-2. The story of paint pigments and their contribution to the development of modern chemistry. A thoughtful and compelling read.

Barovier, Rosa (Mentasti), *L'Arte Vetraria by Antonio Neri.* Milan: Polifilo, 1980, ISBN 88-7050-404-2 (cloth). Introduction in both English and Italian. This, as far as I know, is the only reprint which is a direct verbatim facsimile of an original 1612 printing. It is out of print but still available through some Italian book dealers.

Burckhardt, Jacob, *The Civilization of the Renaissance in Italy.* New York: Barnes and Noble, 1999, ISBN 0-7607-1545-9. Reprint of the 1860 classic. This is ground zero for English language history of Renaissance Italy, a must for anyone seriously interested in the subject.

Catholic University of America, *New Catholic Encyclopedia*, Farmington Hills, MI: Gale Group, 2002. Weighing in at 15 volumes, this is out of range for most individuals, but not university libraries. The original 1913 edition is quite informative and on the web at www.newadvent.org (not affiliated with Gale).

Coyne, G. V., et al. *Gregorian Reform of the Calendar.* Rome: Specola Vaticana, 1983. This is the definitive work on the subject; proceedings of the Vatican conference commemorating 400 years of the Gregorian system. Published in English and Italian versions.

Florio, John, *A Worlde of Wordes*, New York: Gerog Olms Verlag, 1972, ISBN 3487042274, (ed: Bernhard Fabian, et al.) Reprint of the 1598 Italian to English dictionary, out of print and hard to find, but all 480 pages may be downloaded for free from the French National Library website: www.bnf.fr

Freedberg, David, *The Eye of the Lynx.* Chicago, London: University of Chicago Press, 2002. ISBN 0-226-26147-6 (cloth). The story of Prince Fredrico Cesi, and the formation in early 17th century Italy, of what is arguably the first scientific society. This book is a masterpiece.

Garzanti, (Battaglia, Salvatore, et al.): *Grande dizionario della lingua italiana.* Torin: Garzanti, 1961- <in progress>. Not yet complete, this multi-volume dictionary is the Italian counterpart of the OED. 21 vols published as of 2002. While out of reach most individuals, abridged versions are available on CD, or for free on the website: www.garzanti.it.

Hale, J. R., *Florence and the Medici.* London: Phoenix Press, 1977. ISBN 1-84212-456-0 (paper). A popular account of the Medici dynasty by a great historian.

Hibbert, Christopher, *The Rise and Fall of the House of the Medici*. New York: Morrow Quill, 1974, ISBN 0-688-05339-4 (paper). Solid, detailed popular history of the Medicis.

Merrett, Christopher, *The Art of Glass by Antonio Neri*. New York: University Microfilm International, (printed on demand), dist. by www.astrologos.com. This is a direct facsimile print from microfilm of the 1663 translation, owned by Yale University Library.

Sobel, Dava, *Galileo's Daughter*. New York: Walker and Company, 1999. ISBN 0-8027-1343-2 (cloth) Well written account of early 17th century Florentine life through letters to Galileo from his daughter Maria Celeste.

Zecchin, Luigi, *Vetro e Vetrai di Murano*, (3 Vol). Venice: Arsenale, Vol I: ISBN 88-7743-022-2 (1987 cloth), Vol II: ISBN 88-7743-048-6 (1989 cloth), Vol III: ISBN 88-7743-087-7 (1990 cloth). In his lifetime Zecchin became probably the worlds leading authority on the history of Italian glassmaking. These volumes contain his collected work as published in hundreds of periodicals. His writing exhibits the sharp whit, and skepticism which is so essential for accurate historical research. Sadly out of print, and increasingly hard to find the three volumes together, it is definitely worth the effort. (In Italian, with a few English Appendices.)

Zupko, Ronald, *Italian Weights & Measures*. Philadelphia: American Philosophical Society, 1981, ISBN 0-87169-145-0. The final word on a very confusing and complex subject.

PICTURE CREDITS

Front Cover: Buontalenti, Bernardo (1536 - 1608). The main entrance to the Medici 'Casino' Palace, in Florence, demolished in the 19th century, it once stood between Saint Mark's Square and Via San Gallo. Buontalenti himself later made glass there. Credit: Scala/Art Resource, NY.

Back Cover: Full View of the Casino, Credit: Istituto e Museuo di Storia della Scienza, Florence.

APPENDIX A

LETTERE A PRETE NERI
(Letters to Priest Neri)

By Luigi Zecchin
(Vetro e Silicati, Jan-Feb 1964, p17-20)
Translated by Paul Engle

If it is easy to discuss the famous work of Florentine priest-glassmaker Antonio Neri[1], it is not so easy to speak about his life. His first translator, Briton Christopher Merrett, writing in 1662 reported being unsuccessful in finding out anything about him. This was still within fifty years of Neri's death. Nothing was added in the 1817 edition by Milanese editor Giovanni Silvestri who, introducing the last Italian reprint of *The Art of Glass* (*L'Arte Vetraria*), acknowledged being unable to supply "any information about our author".

Little more is revealed by the compilers of a series of biographies written in the second half of 18th and beginning of the 19th century. This was before later volumes about him filled the library shelves. They did not find (or perhaps did not even look for) actual documentation; what little there is here appears to be drawn from their imaginations. The well known *Biographie Universelle*, vol. 31 (Paris, 1822), only says that Neri

"...was born in Florence toward the middle of the sixteenth century. He was ordained by the church, but repeatedly turned down offers of employment and assistance with the aim of being able to completely satisfy his talents for the so-called 'hidden' sciences. He visited the greater part of Europe, stopping in the main cities, and lived for a long time in Antwerp. In many places, he worked in chemical laboratories, posing as a simple technician when other means did not elicit their secrets. In this way he witnessed a multitude of experiments, which he then was inclined to share with the public..."[2]

That, if anything, is vague and inexact. Forty years later, taking greater care, C.J. Poggendorff[3] limited himself to remembering Neri as a Florentine priest who had traveled in Italy and the Netherlands stopping in Murano, and for about a year (around 1609) in Antwerp, then returned to his native land living in Pisa and Florence. To apply some modest extrapolation, all this information is drawn directly from *The Art of Glass*. Indeed, upon re-reading it, just how much more can be learned becomes evident:

"Priest Antonio Neri, having spent many years of his youth working around the glassmaking craft, made a number of his own discoveries in that field. He verified additional techniques provided by men skilled in the art and was induced to compile a treatise in 1612, dedicating it to Don Antonio Medici, his singular benefactor. In 1601, in Florence at the glass furnace of the Casino, he composed batches of a glass paste imitating chalcedony. The scheduling of work at the furnace was the responsibility of his close friend Mr. Nicolò Landi, a talented enamel worker at the oil lamp. Here, the chalcedony presented new opportunities; he made many hundreds of crowns for the gentry, was highly praised by Grand Duke Ferdinando, of blessed memory, and by many other Princes and Gentlemen. In 1602, still at the Casino in Florence, he prepared many batches of aquamarine colored glass for beadmaking cane.

At Pisa, he gained experience with the ash of fern plants for use in the melt. He also made a small furnace, in the form of a frit kiln. Neri regularly manufactured very beautiful aquamarine, and emerald green, and a wonderful pimpernel green for those jobs more exquisite than ordinary, and the colors of sapphire, and transparent red for enameling. In 1602, still in Pisa, he prepared quite a few batches of a blue magpie[T1] colored glass.

In Antwerp, city of Belgium, in 1609 (as a guest for several years of Emanuel Ximenes, Knight of the, renowned Religion of Saint Steven,[T2] Portuguese nobleman and citizen of Antwerp, gentle spirit, well versed in every science), he obtained the most beautiful chalcedony of his life in the glass furnace run by Filippo Gridolphi[4], and with some of it had two vessels made for his Excellency the Prince of Orange. But in Flanders he was also interested in other glass compositions, frequently making, for example, a marvelous aquamarine of his own invention. He also dedicated time to the fabrication of glass paste, imitating precious stones, using the extraction method of Isaac Hollandus,[T3] fusing twenty pots of various colors in a small furnace.

Beyond glassmaking, Neri had cultivated the chemical and homeopathic arts, and after years of work in diverse parts of the world, devoted to the subjects of Chemistry and Medicine, he had it in mind to publish, having experimented by himself and with others, with many impressive effects, both reliable and venerable".

The obstacle to giving a second work to the press must have been his death, which Poggendorff indicates happened in 1614. Although the same date also appears in two later Florentine manuscripts of the same century, the *Sepultuario* of Cirri and the *Poligrafo* of Gargani (reporting the death of "Antonio Neri, experimenter and writer" and of "Priest Antonio Neri, expert in the craft of glassmaking" respectively), [5] I do not know if they are from independent sources.

Similarly, I do not know what reliability can be assigned to a strange genealogical record, supplied by another Florentine manuscript,[6] in which Antonio is identified as "glassmaking author, and then priest", and described (along with a brother "Friar" Vincenzo) as the son of Mr. Neri[7]Antonio[T4], and as the father of four sons (Fillippo, Pierantonio, Andrea, and Francisco). This last piece of information, while not being improbable in the absolute sense, does leave us perplexed. However, the paternity indicated by the record is correct, and it is true that he had a brother named Vincenzo. This can be verified by the two baptismal certificates that follow, filed at the Archives of the Opera di Santa Maria del Fiore, in Florence:

"Thursday, March 1st, 1575:[8]

Antonio Lodovicho was born to Mr. Neri Jacopo and Dianora (daughter of the Franchescho's) and residents of Greater San Pier. The time of birth was February 29th, at 3:00. Appearing are Mr. Franchescho of Girolamo Lenzione, a Florentine lawyer, and Godmother Ginevera di Federigho Sassetti".

"...October 20th, 1579: Vincenzo son of Mr. Neri, Physician, resident of Greater San Pier, was born at 12:30. Appearing is Mr. Piero of the Stifa, Florentine canon lawyer".[9]

These documents, as they appear, not only supply the birth date of the author of *The Art of Glass* (February 29th, 1576 [T5]), the paternal name (Neri), and that of the grandfather (Jacopo, not Antonio as in the genealogical record), but they also inform us (with the baptism certificate of his brother Vincenzo) that the father's profession is that of practicing "physician". Perhaps it was not so strange after all for Antonio to demonstrate a proclivity for medicine.[10]

****

If these bare personal facts do not let us flesh-out the figure of Neri, they at least serve to frame him better in time, and allow us to correct the picture that the old investigators left us, which was further confused by the biographical compendiums of the 1800's. To correct some other misinformation from those same sources, we can turn instead for help to a small file of manuscripts in the National Library of Florence,[11] a little-known collection of 28 letters sent to Neri in Florence and Pisa, starting in 1601. Of them, the first 26, written from Antwerp between August 17th, 1601 and October 31st, 1603 are of particular interest, all from Emanuel Ximenes, who hosted his long visit in Flanders.

On a July 1601, trip to Florence, Ximenes visited his sister Beatrice, wife of Alamanno Bartolini. He stayed with them in the palace that still stands today, at San Trinita Square. This is where he first met Neri, whose duties here are unknown to me; possibly secretary or house master.

His conversations with the young clergyman[12] not only left him with the immediate impression of an exceptional talent, but also of an outstanding guide for instruction in the sciences of chemistry and medicine, for which he also harbored a passion. He was therefore sorry when obligations forced him to leave and endure the return trip to Flanders, unable to enjoy more of Neri's company. But some time later, upon reaching Antwerp and settling in, he realized that through the contacts with his brother-in-law's family he could maintain the correspondence with Neri, and began a dialogue on the subjects that were so close to his heart. In the

first letter, dated August 17th "To the quite magnificent clergyman Mr. Antonio Neri, in the house of Mr. Alamanno Bartolini, in Florence, or where found". After quickly supplying the details of his travels and his health he sends regards to his Florentine "relatives... and in particular to Beatrice", his sister. Here Ximenes lingers to express his great pleasure on receiving a "booklet" of "recipes" from Neri, and declares him "molto caro" (most dear). "With your permission, I will not fail to bother you with my tiresome letters", he then wrote. And the permission would not be denied, indeed, correspondence between the two grew assiduously, and continued for years, extending to the most obscure fields of knowledge.

The topics relevant to glassmaking technology are not lacking in these letters, but they do not appear immediately as we might expect, given the affirmation Neri makes (see chapter 42 of his book) to have made more "pots of chalcedony... in 1601, in Florence, at the glass furnace of the Casino". The only time glassmaking comes up that year is with the news sent by Ximenes on October 26th ("I also have the touch to extract the salt of the bean plant for crystalline glasses..."). That passage reports of a personal investigation by the Belgian,[T6] which lacks the expressions of gratitude, and the assurances to take good notes, which will later accompany the reception of new procedures sent by the Florentine priest ("I have received the recipe for the color turquoise... I give you my heart-felt thanks... Interim I have noticed the way prescribed in the book of recipes from Your Holiness", wrote Ximenes on November 15th, 1601; and on March 28th, 1602: "Also I have noted your melt recipe with iron that has a gold cast, and for that I thank you, Your Holiness").

We must wait a full year in order to find a sign of Neri's own glassmaking experiences: it is in a letter of October 31st, 1602, in which Ximenes thanks him for some glasses sent from Florence, remarking "how very beautiful your variety is". These were probably examples of the chalcedony for which he would soon receive the precise recipe.

The absence of any glassmaking discussion in the first letters from Antwerp, the unexpected reference appearing in October 1602, and the persistent allusions in subsequent letters, leads to the conclusion that in 1602 (and not in 1601) Neri began to turn his attention to the problems of glass fabrication.T7 What circumstances accelerated his interest, I just do not know. It might have been Bartolini that introduced him to the laboratories of the 'Casino' where Don Antonio Medici, his contemporary, gave him free rein to pursue his passion for alchemy.13 It is thought that Neri, after some time,

was put into service as an active glassmaker at the "Casino",14 feeding his taste for research and practical investigation of technique. At this phase, he worked independently, initially turning his attention to the more visually pleasing glass, the 'chalcedony'.

He precisely imparts to Ximenes the "detailed recipe" of this glass, writing him on November 9th, 1602; and in the same letter, he expresses the desire to visit, in order to work together a bit. And Ximenes, after thanking him for the recipe and for answering the profusion of questions on the details of execution, continued effusively on December 5th: "I have seen the tender affection which Your Holiness shows me, and demonstrates with the hope to see me before death, which is no different from my own hope. I have desired this from the start... because if we were together, we could easily set to work on some small projects, being that our talents, if I am not deceiving myself, are very well suited...". And he added that the surer way of arriving in Flanders, for Neri who did not speak "German or Flemish", would be to go by way of Venice, and to join with the merchants on their way to the "fair held in Frankfurt during Mid-Lent", and at its conclusion, continue on to Antwerp with the merchants returning there. The good results obtained with the chalcedony encouraged Neri to extend his tests to pastes for beadmaking cane, and supervised their production, not only in Florence, but also in Pisa[15].

Ximenes had all of the glassmaking activity in that city related to him by Neri, in a letter of November 19th. The Belgian confirmed his constant "desire to see you", and assured him that he would attend to arrangements and would supply him with money for travel expenses, writing on December 13th: "I feel glad that Your Holiness has had such an influence in the glassmaking craft. Is it not wonderful? It is certainly an extraordinary ability, to have such sharp powers of investigation and keen eyes.[T8] While I eagerly await what life will bring, you hear and see new opportunities..." And again on February 7th, 1603: "I feel glad that Your Holiness is having such success with enamels and glasses..."

Neri became so absorbed in glassmaking research at the furnaces in Pisa, that he became ill and had to postpone the planned travels to Flanders. "Praise God that your indisposition has ended - Ximenes wrote affectionately on May 2nd -"... if the Pisan air is suited to your recovery do not change it, because health is more important than wealth. As far as coming to these parts or not, Your Holiness, simply trust in God..." But the Belgian was also infected by the passion for the glass, and was by then experimenting with the fusion of rock crystal: "I am sure *that* will succeed - he reported - Because I have already demonstrated it here, mixing in the bean plant salt, it really worked very well, and was much better when made in the kiln. It likes being in the frit, which you tested with the iron rake[T9] (in fact, I think [the salt] sticks to the paste quite well and I will send you what remains of the batch bonded with zaffer[T10]), since then I have not made more ..."[T11]

Meanwhile, Neri resumed reporting the further details of results obtained in his new tests. "As far as the red glass - wrote Ximenes on June 13th of the suitability of an enamel he had ordered from Pisa, purposefully overloaded with color - I am doing a test in enameling gold, because having such a thin layer of enamel, if it is not very full of pigment it would remain a pale color... If it will not grieve Your Holiness, I would love to find out the composition... ".

And on June 27th: "It would give me great pleasure to have Your Holiness' new and improved chalcedony. To me this chalcedony really seems to be something of an improvement. To me it really seems to be a little more natural, the confirmation of which we saw with the gentry as your grace wrote to me some days ago, that it astonished all the officials, as well as goldsmiths and glassmakers alike. That being the case, when you get a chance I hope to see a little of the new..." But the long awaited sample was still not received by July 11th, when the Belgian wrote: "The details of the last chalcedony, which you promised to send to me, did not come in the letter: but I had to recant by the time I got to the end... I see, and understand, than Your Holiness it is not in leisure, but in fact busy at work in the service of Christianity..."

Amidst "working in the service of Christianity" we know Neri also devised "the way to make vegetable salt from ferns, that makes a very beautiful crystal", recorded in chapter 5 of *The Art of Glass*. He had informed Ximenes in late May of the research he proposed to do, and must have followed with positive results, because on October 31st 1603 the Belgian writes "I had anticipated a long wait for the secret of the ferns..."

It would seem that Neri began his travel towards the north then, or soon after: with the last letter dated October 1603, in any event, the correspondence of Emanuel Ximenes pauses. Resumption came with the 27th letter of our bundle, which precisely fixes the date that the glassmaker priest returned to his native land: "Antwerp, last of March 1611", in fact he thanks the "Very Reverend Father Antonio Neri, in Florence", for the "three lovely deliveries of February 3rd, 12th, and 26th", which were received "after your departure."

There is still more to be learned from Ximenes' letters, which I will leave for another time. At this point I want to propose a realistic biographical sketch: Born in 1576, ordained clergyman by 1601, Neri began his glassmaking activity in the first half of 1602, and he practiced it in Florence and Pisa towards the end of the successive year. At the beginning of 1604 he moved to Antwerp where he stayed until January of 1611, and then returned to his native land. Here he reordered his notes on glassmaking, entrusting them to the press in January 1612, and in 1614 (presumably) he died.

This is an outline that has some gaps: but none wide enough to contain the troubling error of a "ramble through the greater part of Europe stopping in the main cities", and the secretive "entrées to chemical laboratories posing as a simple technician", stories which have been repeated for ages and used to base the biographies of Reverend Neri.

Footnotes:

[1] Cf. L. Zecchin.: *Il libro di Prete Neri*, [The Book of Priest Neri], Vetro e Silicati, November-December 1963.

[2] Op. sit. p.247 of vol XL of the contemporary Italian translation in 65 volumes, edited by G.B. Missiaglia in Venice, between 1822 and 1831, with the title: *Biografia Universale antica e moderna, ossia Storia per alfabeto della vita publica e privata di tutte le persone che si distinsero oer opere, azioni, talenti, virtù e delitti. Opera affatto nuova compilata in Francia da una con aggiunte e correzioni.* [Ancient and Modern Universal Biography, An Alphabetical History of the Public and Private Lives of All Persons of Distinguished Works, Actions, Talents, Virtues and Crimes. Newly Compiled in France with Addendum and Corrections.]

[3] *Biographisch-Literarisches Handwörterbuch* [Biographic-Literary Hand Dictionary] (Leipzig, 1863); part 2a, col. 270.

[4] About Filippo Gridolphi: *la personnalité dominante de l'art du verre dans les Pays-Bas*, [Dominant Personalities in the Art of Glassmaking in the Netherlands], V.R. Chambon: *Histoire de verrerie en Belgique, du 11.me siècle à nos jours.* [The History of Glassmaking in Belgium, 2nd Century to the Present], Brussels, 1955.

[5] National Library of Florence: Sala manoscritti [Hall of Manuscripts].

[6] State Archives of Florence: Pucci, Alberi Genealogici [Family Trees].

[7] Neri, is a shortened version of the name Ranieri.

[8] The year is indicated using the old Florentine calendar, and corresponds to 1576.

[9] I must credit the transcription of the two certificates (and the information found in the previously cited Florentine manuscripts) to Mr. Enzo Settesoldi, of the Opera del Duomo [the records office of the Santa Maria del Fiore Cathedral (Duomo), in Florence]. From the same Mr. Settesoldi I learned the date of the wedding of Antonio's parents (10 August 1570) and the dates of birth of the other siblings, Jacopo, Francesco, and yet another Jacopo (12 Decembers 1573, 16 February 1575, 30 May 1577 respectively).

[10] To remove any doubt of the father's profession, the baptismal certificate of the brother Francisco bears the indication "... di M.o Neri di Jacopo Neri, phisico, et di Dianora de Parenti..." ["... of Mr. Neri (son of Jacopo Neri), Physician, and of Dianora; the parents..."].

[11] Ms. II, I, 391 (Magl. Cl. [sec.] XVI, n. 116).

[12] The date of Neri's ordination to the church is not known (The archives of the Archdiocese of Florence currently have conservation efforts under way to restore surviving ordination records dating back to 1650, previous ones were destroyed by a fire); but his status in the church is indicated in the letters of 1601. I have found no confirmation of his title being monk or abbot, as assigned by some authors.

[13] Don Antonio (1576-1621), the son of commoners but passed as the offspring of Bianca Cappello and Francisco de Medici (Grand Duke of Tuscany from 1564 to 1587), was raised like a prince in the Medici household and treated as such even after the truth of his origin was known (but was conferred with the cross of the Knights of Malta,T12 in order to ensure his celibacy). He made his residence in Florence, at the Medici palace called the 'Casino' located between Saint Mark's Square and Via San Gallo. There he was trained by alchemists, spending, as one historian from Tuscany wrote two centuries ago, "immense sums of gold in order to learn and to experience various secrets that were sold to him by imposters at a dear price", but succeeding also "to collect and to verify a great number of secrets pertaining to medicine, and to the perfection of diverse arts" (cf. Notize de G. Taragoni Tozzetti; Florence, 1852 p 256).

[14] Grand Duke Francisco had it build around 1575.

[15] The glasshouse in Pisa, originally constructed by Cosimo I, Francisco's father, had been renovated by their successor, Fernando (Grand Duke from 1587 to 1609).

Translators Notes

[T1] 'Gazzera marina' it would seem refers to a species of birds familiar to Neri, but not to us, or at least not familiar by that name. Its color is distinctly 'celestial' blue as evidenced by the title of chapter 23 of *The Art of Glass*: "Celeste colore o vera gazzera marina vetro". The obvious candidate, 'gazza marina', or razorbill, is a relative of the penguin and does frequent the Italian coast, but sadly sports only black and white plumage. Gazza or gazzera also refers to the magpie, and if 'marina' is taken to mean blue, in the same sense that Neri uses 'acqua marina' then our attention is drawn to the Iberian '*gazza azzurra*' [blue magpie = cyanopica cyanus], which is to my mind a credible fit, although there is certainly room for debate.

[T2] The Sacred Order of the Knights of Saint Steven, a Catholic religious order of seafaring crusaders, founded by Grand Duke Cosimo Medici in 1562, as a Tuscan coast guard to protect the shores of the Tyrrhenian Sea, mainly from Turkish pirates.

[T3] "Isach Olando" = Isaac Hollandus

[T4] The paternity is correct in that his father was '*Neri*'. Other records [9], [10] show his grandfather as '*Jacopo*' and not '*Antonio*', thus Neri di Jacopo rather than Neri di Antonio.

[T5] Besides the correction for new years day [8] there were 10 days deleted from October in 1582 by order of Pope Gregory XIII. On our calendar Neri's birthday falls on 10 March 1576.

[T6] I have substituted '*Belgian*' for the less mellifluent '*Antwerpan*'.

[T7] Zecchin sees a contradiction of dates where there may not actually be one. Neri states clearly that he grew up around glassmaking (an art critical to his father's profession), and Ximenes refers to glass recipes in the booklet he received from Neri before the first letter in 1601. It seems plausible that Neri was involved with glass all along, and it was Ximenes' interest that really picked up in 1602.

[T8] Literally '*eyes of a lynx*'. In classical Greek mythology Lynceus was the grandson of Perseus, and had preternaturally keen eyesight. See Apollodorus, *Bibliotheke* I, viii, 2 & ix, 16; III, x, 3 & ix, 2. Coincidently it was Prince Federico Cesi, the founder of the scientific society *Accademia dei Lincei* [Society of the Lynx-eyed] that brought Neri's book to the attention of Galileo Galilei in 1614.

[T9] See *A Worlde of Wordes*, J. Florio, pub A. Hatfield, London 1598, p459. "Zaffara -
An iron hook or drag used by dyers..."

[T10] See *Vetro E Vetrai Di Murano*, L. Zecchin vol.3 pub Arsenale Editrice, Venezia
1990, ISBN: 88-7743-087-7, p18. [Zaffera; a mixture of silica sand and cobalt oxide
produced in Saxony]. Also *Materials Handbook*, 12th ed. G. Bradey et.al. Pub. Mcgraw-
Hill 1986, ISBN: 0-07-007071-7 p204: "...a deep-blue powder made by fusing cobalt
oxide with silica and potassium carbonate. It contains 65 to 71% silica, 16 to 21potash,
6 to 7 cobalt oxide, and a little alumina."

[T11] My translation of this passage is suspect at best, the original reads: "...per invidia,
stando in fritta, vi missono della zaffara (anzi, credo che pigliassono la pasta buona e mi
mandassino altra cattiva inzafferata), onde qui non ne ò fatto altro..."

[T12] Knights of Malta, like the Knights of Saint Steven [T2] were seafaring crusaders,
coastguards, and slave traders, but by 1600 largely corrupt, run by and for the benefit of
the aristocracy. Conferees took a vow of celibacy and poverty.

BIOGRAPHICAL TIMELINE BASED ON THE ARTICLE

Dates *	Events
1575 Feb 29	Antonio Lodovicho Neri was born to Dr. Neri Jacopo and Dianora.
1579 Oct 20	Vincenzo son of Dr. Neri, Physician, resident of Greater Saint Pier, was born.
1582 Oct 4	10 days deleted from the calendar by decree of Pope Gregory XIII, tomorrow is October 15th
1585-1600	Neri comes of age and is ordained Priest.
1601	In Florence he composed batches of a glass paste imitating chalcedony (chapter 42).
1601 July	Emanuel Ximenes visited his sister in Florence, and meets Neri for the first time.
1601 Aug 17	In the first letter from Antwerp Ximenes thanks Neri for booklet of recipes.
1601 Oct 26	The first time glassmaking comes up "...salt of the fava bean for crystalline glasses..."
1601 Nov 15	Ximenes: "I have received the recipe for the color turquoise... I give you my heart-felt thanks".
1602	Still at the Casino in Florence, Neri prepared many batches of glass for cane used in beadmaking, as well as of aquamarine colored glass Zecchin postulates that this is the actual year Neri turns his attention to glass fabrication.
1602	At Pisa, Neri gained experience in fusing the ash of fern plants for use in the melt. He also made a small furnace, in the form of a frit kiln. He regularly manufactured very beautiful aquamarine, emerald green, and a wonderful pimpernel green, for those jobs more exquisite than ordinary, and the colors of sapphire, and transparent red for enameling.
1602	Still in Pisa, he prepared quite a few batches of a blue magpie[T1] (sky) colored glass.
1602 Mar 28	Ximenes: "also I have noted the recipe for the iron melt that casts like gold".
1602 Oct 31	First sign of Neri's own glassmaking: Ximenes thanks him for samples, and notes its beauty.
1602 Nov 9	Neri sends a detailed glass recipe, and expresses the desire to visit Ximenes in Antwerp.

Dates*	Events
1602 Nov 19	Neri relates the glassmaking activity in Pisa, Ximenes, assures him that he would attend to arrangements and would supply him with travel expenses.
1602 Dec 5	And Ximenes, after thanking him for the recipe and for answering the profusion of questions on the details of execution, continued effusively "I have seen the tender affection which Your Holiness shows me, with the hope to see me before death..."
1602 Dec 13	"I feel glad that Your Holiness has had such an influence in the glassmaking craft. Is it not wonderful? It is certainly an extraordinary ability, to have such sharp powers of investigation and keen eyes. While I eagerly await what life will bring, you on the other hand, hear and see new opportunities..."
1603 Feb 7	"I feel glad that Your Holiness is having such success with enamels and glasses..."
1603 Mar	Neri became so absorbed in the glassmakers research at the furnace in Pisa, that he became ill and had to postpone the planned travel to Flanders.
1603 May 2	"Praise God that your indisposition has ended ... if the Pisan air is suited to your recovery do not change it, with health passes riches... As far as coming to these parts or not, Your Holiness, simply trust in God... "
1603 Late May	Neri informs Ximenes of research he proposed to do (ferns).
1603 Jun 13	"As far as the red glass - wrote Ximenes on June 13th of the suitability of an enamel he had ordered from Pisa, purposefully overloaded with color - I am doing a test in enameling gold, because having such a thin layer of enamel, if it is not very full of pigment it would remain a pale color..."
1603 Jun 27	"It would give me great pleasure to have Your Holiness' new and improved chalcedony. To me this chalcedony really seems to be something of an improvement. To me it really seems to be a little more natural, the confirmation of which we saw in the gentry as your grace wrote to me some days ago, that it astonished all the officials, as well as goldsmiths and glassmakers alike... Never the less I will patiently wait to see a little of the new one... "
1603 Jul 11	" The details of the last chalcedony, which you promised to send to me, did not come in the letter: ... I see, and understand, than Your Holiness it is not in leisure, but in fact busy at work in the service of Christianity... ".
1603 Oct 31	I had anticipated a long wait for the secret of the ferns... This is the last letter from Ximenes until after Neri returned to Italy.

Dates *	Events
1609	In Antwerp he obtained the most beautiful chalcedony of his life in the glass furnace run by Filippo Ghiridolfi, and with it had two vessels made for his Excellency the Prince of Orange.
1611 Feb3, 12,26	Ximenes receives three parcels from Neri after his departure from Antwerp.
1611 March 31	Ximenes 27th letter addressed to Neri in Florence thanking him "for the "three lovely deliveries of February 3rd, 12th, and 26th", which were received "after your departure."
Jan 1612	Neri delivers manuscript of The Art of Glass to the printers.
1614	The obstacle to giving a second work to the press must have been his death.

* Dates are as recorded on the documents at the time.

APPENDIX B

Editions of The Art of Glass

Year	Lang. Ed/Imp.	Title	Place of Issue/Pub	Subject Author/Translator
1612	Italian 1st	L'Arte Vetraria	Florence Giunti	Original Antonio Neri
1661	Italian 1st /2nd	L'Arte Vetraria	Florence Fortuna	Original "Corrected" Marco Rabbuiati
1662	English 1st	The Art of Glass	London A.W. for Octavian Pulleyn'.	Orig. Trans. & Appended C.M. (Christopher Merrett)
1663	Italian 2nd	L'Arte Vetraria	Venice	Original modified Giacomo Batti
1668	Latin 1st	...de Arte Vitraria	Amsterdam	Trans. of Neri-Merrett Andreas(m) Frisius(m)
1669	Latin 1st /2nd	...de Arte Vitraria	Amsterdam	Trans. of Neri-Merrett Andreas Frisius (Frisium)
1670	Latin 1st /3rd	...de Arte Vitraria	Amsterdam	Trans. of Neri-Merrett Andreas Frisius
1678	Italian 3rd	L'Arte Vetraria	Venice	Original modified Steffano Curti
1678	German 1st	L'Arte Vetraria	Frankfurt & Leipzig	Trans. of Neri-Merrett Friedrich Geissler
1679	German 1st/2nd ?	Ars Vitraria Experimentalis...	Amsterdam & Danzig	Trans. of Neri-Merrett & App. Johann Kunckel von Löwenstern
1679	German 1st/2nd ?	Ars Vitraria Experimentalis...	Frankfurt & Leipzig	auff Kosten des Autoris Trans. of Neri-Merrett & App. Johann Kunckel von Löwenstern
1681 ?	Latin 2nd ?	...de Arte Vitraria	Amsterdam	Trans. of Neri-Merrett Andreas Frisius
1686	Latin 3rd (2nd?)	...de ArteVitraria	Amsterdam Henr. Wetstenium?	Trans. of Neri-Merrett Andreas Frisius?

Year	Lang. Ed/Imp.	Title	Place of Issue/Pub	Subject Author/Translator
1689	German 2nd	Ars vitraria experimentalis ...	Frankfurt & Leipzig Riegels	Trans. of Neri-Merrett & app. J. Kunckel
1697	French 1st	Art de la Verrerie	Paris	Plagiarized Trn. of Frisius Jean Haudicquer de Blancourt
1699	English	Art de la Verrerie	London Printed for Dan. Brown ... Tho. Bennet ... D. Midwinter and Tho. Leigh ... and R. Wilkin ...	Trans. of H. Blancourt
1743	German 3rd	L'Arte Vetraria	Nurnberg In verlegung C. Riegels	Trns. of Neri-Merrett & app. J. Kunckel
1752	French 1st	Art de la Verrerie	Paris Durand Pissot.	Trns. of Neri-Merrett-Knuckel M.D. *** (Paul-Henry Thiry, Barron of Holbach)
1756 1765?	German 4th	L'Arte Vetraria	Nurnberg	Trns. of Neri-Merrett & app. J. Kunckel
1759	French 2nd	Art de la Verrerie	Paris	Holbach
1776 1778?	Spanish 1st	Arte de Vederia	?	Dom Miguel Geroni[m/c]o [S/G]uarez
1781	Italian	L'Arte Vetraria	Venice	Francesco di Niccolo Pezzana
1785	German 5th	L'Arte Vetraria	Nurnberg	Trns. if Neri-Merrett & app. by J. Kunckel
1787?	Italian	L'Arte Vetraria	?	
1817	Italian	L'Arte Vetraria	Milan	Antonio Neri Giovanni Silvestri
1826	English 2nd	L'Arte Vetraria	Middle Hill	Neri-Merrett reprint ed: Sir Thomas Phillips Barton

Year	Lang. Ed/Imp.	Title	Place of Issue/Pub	Subject Author/Translator
1975	German	Ars Vitraria Experimentalis...	Volkseigener Aussenhandelsbetrieb der DDR	Reprint of 1679 ed.
1980	Italian	L'Arte Vetraria	Milan Edizoni il Polifino	Reprint of original verbatim. Rosa Barovier Mentasti ISBN 88-7050-404-2
1992	German	Ars Vitraria Experimentalis...	New York Georg Olms Verlag	Reprint of 1689 ed. of Neri-Merrett-Kunckel
2001	Italian	L'Arte Vetraria	Florence Giunti Gruppo Editoriale	Reprint of original (updated typography) Ferdinando Abbri ISBN 88-09-01267-4
2001	English	The Art of Glass	Sheffield Society of Glass Technology	Reprint of Neri-Merrett ISBN 090068237X
(2002)	English	The Art of Glass	New York UMI Books on Demand (astrologos.org)	Reprint from microfilm copy of Yale University Libr. (verbatim) copy of Neri Merrett

NOTES:

1. Considering that all non-Italian versions (present work excluded) derive from Merrett's 1663 English work, and many from the Latin translation of Merrett, it is not surprising that all contain errors, some considerable. What is surprising, even astounding, is that some later versions bear any resemblance to the original at all! Consider the 1699 English volume (Brown et.al.), published only 36 years after Merrett, tracing a tortured lineage of Italian (original), to English (Merrett), to Latin (Frisius), to French (Blancourt), back to English again.

2. While considerable time and effort have been expended on this listing, it should under no circumstances be considered 'complete and fully correct'. It represents the best efforts of the editor to date, but almost certainly contains some errors and omissions. For example, the 1686 Latin edition is almost certainly a mis-cataloging of the 1681, but the resources to rule it as such have not been available. Also the 150 year gap covering the mid 19th and 20th centuries demands greater diligence. It is hoped that corrected information will be forthcoming for inclusion in the hardcover edition.

"The moon should be waxing, and close to its opposition with the sun, because at this point the plant is in its perfection, and gives a lot of salt, more than it would at other times, and of better nature, strength, and whiteness."

- Antonio Neri

APPENDIX C

Weights, Measures & Miscellany
of the 16th - 17th Century

Varied Between Towns and Regions:
(Used in Florence, Pisa, Rome, Volterra)
1 Roman Pound (Libbra) = 12 Onces = 339 grams
1 Roman Ounce (Once) = 1/12 Libra = 28.25 grams

In Liguria 1 Libbra = 317 grams
In Massa 1 Libbra = 332 grams
In Piedmont 1 Libbra = 307 grams

1 Modern (avoirdupois) Pound = 16 ounces = 453.6 grams
1 Modern (avoirdupois) Ounce = 1/16 Pound = 28.35 grams

Florentine New Years day: March 25th
By Papal decree, the day after Thursday, 4 October, 1582 was Friday, 15 October.
Leap Years: before 1582, all years divisible by 4, but using January 1 as new years day;
After 1582 same rules as today.

Dates before 15 October, 1582 + 10 days = Those dates on our current calendar
Dates on our calendar before 15 October, 1582 - 10 days = As they would be written then

Currency:
1 Florin = approx. 28mm dia.

Neri's Hierarchy of Glasses:

Common Glass - (*vetro commune*) - the lowest quality glass
Crystallino - (*cristallino*) - medium quality, between common and crystal
Crystal - (*cristallo*) - high quality glass, exceptionally clear and workable
Artificial Rock Crystal (*bollito, cristallo [di montagna] artificiale*) - finest crystal glass

"In closing, I say that the artisan who is diligent, practical, and works step by step, as I describe, will find truth in the present work."

- Antonio Neri

APPENDIX D

Antonio Neri's Birthday:

10 March 1576

According to a baptismal certificate filed at the Opera de Santa Maria del Fiore, in Florence, Italy; on Wednesday, 29 February 1575 at 3:00, Dianora and Neri Jacopo became the proud parents of a bouncing baby boy named Antonio Lodovicho. Later centuries would remember him as Antonio Neri, author of perhaps the most celebrated book in the history of glass making, *L'Arte Vetraria*, or *The Art of Glass*.

Now, four hundred odd years later, contemporary glass workers and artisans conceivably might want to mark their calendars, and when the appointed day arrives, raise a glass in celebration, and drink a toast to the esteemed glassmaker-priest. However, upon closer examination of the actual birth date on the certificate, some puzzling questions arise. Most notably, 29 February, at least on my calendar, occurs only in leap years, and leap years are always even-numbered. Could 1575 somehow have truly been a leap year in Florence? In fact it was. Florentines, and indeed many Renaissance Europeans rang in the New Year with the rebirth of the natural world at the beginning of spring,

on 25 March. Leap years, as well as the dates of Easter Sunday, and other holy observances were set by another system; the ecclesiastical calendar. This official church calendar started the New Year as we do today, on 1 January.

Correcting for the New Years day discrepancy, we arrive at Wednesday, 29 February 1576, which sounds more plausible as the date of Neri's birth. But again, closer scrutiny reveals that by our current calendar that date should have fallen on a Sunday, not Wednesday. The explanation is a bit more involved, but stay with me. Again, the solution involves leap years. Adding a day every fourth year is a practice that originated back in the Roman Empire, and while some early mistakes were made, the scheme was faithfully observed from then on. A problem with that system was that the average length of a calendar year was 365¼ days, but the earth completes its orbit around the sun in slightly less time. The result is that spring arrived earlier and earlier; by about 11 minutes each year, or a full day every 128 years. By the time Antonio Neri was born, spring was beginning in early March (astronomically speaking) and the calculations for Easter were seriously messed up. The solution, decreed by Pope Gregory XIII in 1582 was as follows. First, years evenly divisible by 100 would hence forth not be leap years unless also divisible by 400. This reduces the error to about one day every 3300 years. Second, the day after Thursday, 4 October 1582 would be Friday, 15 October. This fixed most of the previous slippage in the calendar moving the date of the vernal equinox, and the first day of spring back to 21 March. The easy way to think about this is that 1582 was a special year, ten days short, which is to say only 355 days long.

Finally, to correct for 1582 we account for it in our current calendar as if it were a full length normal year. This means adding ten days to dates recorded before the change and therefore, on our calendar the particular Wednesday that Antonio Neri was born would be 10 March 1576! So what are you waiting for? Mark it down, and make plans. What better way to celebrate our extraordinary heritage of glassmaking: Salve Antonio!

1. L. Zecchin, Lettere a Prete Neri, Vetro e Silicati, Jan-Feb 1964, p17-20.
2. A. Neri, L'Arte Vetraria, Florence, pub: 1612 Giunti.
3. G. Coyne, et al, Gregorian Reform of the Calendar, pub: 1983 Vatican.

INDEX